FARM SEASONS

FARM SEASONS

by

Ray Dimmock

FIRST EDITION

Copyright © 2021 Ray Dimmock

All rights are reserved under International and Pan-American Copyright Conventions. Except for brief passages quoted in a newspaper, magazine, radio or television review, no part of this book may be reproduced in any form or by any means, electronic or mechanical, including photocopying and recording, or by any information storage and retrieval system, without permission in writing from the author.

Times New Roman, Minion Pro, Arial, and Gils Sans.

First Edition, 2021, manufactured in USA
1 2 3 4 5 6 7 8 9 10 LSI 25 24 23 22 21

Front cover painting of the two barns is by Mrs Beatrice Dimmock (Mom).

Dedication photograph is of Beatrice & Harold Dimmock (Mom & Dad).

All other photographs and paintings are listed at the back of this book.

Susan Dimmock edited the original manuscript.

ISBN 978-0-578-90782-6 (pbk. : alk. paper)

Contents

Acknowledgements	xi
Introduction by Ray Dimmock	xiii

Chapter 1 - The Farm	1
Chapter 2 - Spring	5
Chapter 3 - Summer	13
Chapter 3A - Siloing	17
Chapter 3B - Haying	21
Chapter 4 - Fall	33
Chapter 5 - Winter	35
Chapter 6 - Managing the Cows	39
Chapter 7 - Milk Processing	43
Chapter 8 - The Milk Route	47
Chapter 9 - Maintaining the Farmhouse	55
Chapter 10 - Fun on the Farm	61
A - Snow Sledding	62
B - Henry's House and 4H Club	62
C - Riding Pete	64
D - Power Snow Sledding	65

E - Chickens	66
F - Summer Evenings	67
G - Lloyd Road Sports	73
H - Motordrome	74
I - Horse Chestnut Battles	74
J - The Secret Room	74
K - Sunday Rides	75
L - Boy Scouts	75
Chapter 11 - Memories Now and Then	77
References	87
List of Photos and Paintings	89
About the Author	92

Dedication

This book is dedicated to Mom and Dad. Their lessons in life were important to all of us, helping to lead each of us, in our own way, to successful lives of our own.

Acknowledgements

This project has been a labor of love, considered initially nearly ten years ago, but completed primarily during the last year. Over these years, I've had many suggestions from my siblings. My role has been to assemble a first draft to be sent out to the sibs for their comments and suggestions. I was incredibly pleased to incorporate those comments and add to the story. It is now about twice the size of the first draft, so clearly, their input was very helpful.

While all their comments were important, one suggestion by sister Mary resulted in adding an entire chapter, the one describing the role of the women in our family. It is embarrassing to me that I needed to be reminded of their part in keeping everything running smoothly.

One of the most important sources of pictures used in the book is a DVD that was transferred from 35mm films taken about 60 years ago. Thanks to brother-in-law Clay for his efforts in creating the DVD; many of the pictures would not have been possible otherwise.

I consulted Wikipedia to clarify information about both the Harkness and Hammond estates and found it to be helpful. I also reviewed *An Illustrated History of Waterford, Connecticut* by Robert Bachman, copyrighted in 1967. It confirmed some of the information about the Booth Brothers quarry. One interesting tidbit from that book was one picture showing several of the cows in our barn.

Many thanks to Susan Dimmock, my niece, for her expertise in editing this story. She ironed out all the grammatic wrinkles to make it a much better document.

Thanks also to Michael Linnard for his very significant assistance, guiding me through the minefield of self-publishing.

I am proud of the story that is portrayed here, and incredibly grateful to all the efforts, suggestions and corrections provided by so many. I know that the book will be enjoyed by those that have an interest in farm life in the middle of the twentieth century.

Introduction

I left our dairy farm in Waterford, CT more than a half century ago. At my age, it is quite common for one's memory to give out, and mine certainly is on its way there. I have, though, always had good recall of numbers—phone numbers and license plates, for example. Even now, I can recall our phone number when it was just a 4-digit number on a party line. My grandchildren do not believe that there was ever a time when there were such short numbers. And, of course, they have no idea what a rotary phone is. I also remember the license plate numbers for our family vehicles from the 50s and 60s.

I retain some mental images from my farm days too that are as clear today as ever. When I was 10 years old, Dad brought home a brand-new milk truck, a 1956 Chevrolet panel truck. It had a 4-speed automatic transmission, which was becoming more common in cars in 1956 but rare in trucks. The truck could be driven either standing up or sitting down, a very convenient feature for milk delivery.

The image I still have of the day my father brought the new truck home is him pulling it into the driveway, parking it in front of our house and then asking me, a 10-year-old, to drive the truck down to our fuel tank to fill the tank with gas. I had been driving farm vehicles for a few years, so driving was not unusual for me, but driving our brand-new truck was special, a memory that is so vivid still. This is one great example of how we were all given responsibilities at an early age.

The truck was a major purchase for the farm, one of a very few new acquisitions. Our previous milk trucks were pickup style, which meant that, at each customer, we had to stop the truck and get out and fill the customer's order from the back of the truck. With the new truck, the milk was inside with us, which meant that the runner could prepare the delivery for the next customer while the driver was en route. Upon arrival at the customer, the delivery could be made immediately, thus shortening the whole route. The runner could even jump out just before the truck came to a complete stop.

Over the years, I have collected many of these memories. On those occasions when I would happen to mention them at family gatherings, my siblings would suggest that I should write a book. I always thought it was a good idea but procrastinated for years. What finally triggered the project a few years ago were several sightings of old farm equipment in my travels, especially near my home here in Ashford.

Less than a mile from home, there is a side-delivery rake right along the road. My daughter Rae showed interest in knowing what it was and how it worked, and of course I loved explaining it. On another road close by is a trip rake, quite like one that we had. At a nearby farm, there is an old hay loader, again like ours.

The book project went dormant from when we first saw the rake, but now I am back at it. One of the reasons is that valuable resources for the book will not be accessible forever. In fact, perhaps the best resource would have been my cousin Don. He died suddenly just a couple years ago. His dad and mine were brothers and owned and ran the farm together. Of all the cousins, Don spent more time working on the farm than the rest of us. Another great resource would have been our oldest sibling, Henry, who passed in 1991.

Growing up on the farm helped make me who I am. There were over 100 farms in the town of Waterford in the 1920s, so lots of people experienced farm life, either directly or at least through indirect exposure. There were at least 6 farms within one mile of our farm when I was young. There are very few farms in Waterford now, so I am hoping that documenting my memories will be enjoyable to those who did not have the experience of farm life, or perhaps worked on a farm in the past. I'm also hoping that my kids and theirs will get a sense of what farm life was like.

Farm Seasons is about the various activities that happened throughout the year on our farm, and I believe that similar activities took place on many farms over the course of the year. I will try to relate the things we did and the equipment we used. I will contrast our methods to current methods where appropriate. While modern farms do things much differently than we did, with new and more

Figure 1 - The Farm House, Hay Barn and Cow Barn.
The Milk House is to the far left of the Farm House

efficient equipment and methods, I'm sure the rhythm of their annual seasons is quite similar to ours.

Summers were always busy on the farm, and the staff included our families and several neighbors.

Participation of my siblings on the farm is divided into four groups. Group one was Henry, Ed and Dot. Group two was Dick and Cousin Donny. Group three was Cousin Wayne and me and represents the primary time discussed in the book. Group four was Al and his friends including Mike Kelley and John Merrill. Since Al was only 9 when I left, the Floyd brothers filled in that gap.

Summers were always busy on the farm, and the staff included our families and several neighbors. As each of the older family members moved on, the younger ones stepped into increased responsibilities. I myself did not appreciate my farm experience nearly as much then as I do now. It was hard work but enjoyable, too.

Chapter 1
The Farm

Figure 2 - The Farm as viewed from the first hill.

Dimmock's Dairy was located on Dimmock Road in Waterford, CT. Our farm comprised approximately 114 acres, spreading from Dimmock Road to the west, almost to Great Neck Road near Great Neck School. To the south was Verkade Nursery and the Anderson Farm. At the far west was the Booth Brothers Quarry, just off Great Neck Road near West Neck Cemetery. To the northwest was the Perkins Farm. To the north was the former Gallagher Farm. While this view does not show our Milk House and the Horse Barn, they were just to the southeast of the Hay Barn, which would be to the right side of the above picture.

The adjacent quarry was famous in its day but lost power during the Hurricane of 1938, which flooded the lowest parts due to loss of dewatering pumps. The resulting deep pond still exists today. When we were kids, we used to go fishing at the quarry from time to time. Back then, we also could see remnants of the quarry operation, including piles of stone and cables that were used to hold up the tall poles used to support stone extraction equipment. It's known that granite from this quarry made its way to buildings in Hartford and New York City.

The farm was started in 1847 with the purchase of 75 acres of land by my great, great grandfather, Rufus Leverett Dimmock. Over the years, additional land was purchased, increasing in size to the 114 acres that we had during the final days of operation. Great-Grandfather Leverett Nye then took over, followed by Grandfather Oscar Rufus and, finally my Dad Harold and his brother Uncle Francis, known to all of us as Ike (pronounced Ikey).

The farm was forced to close in 1972 due to family illness. Uncle Ike became sick, making it impossible for Dad, who also was in failing health, to keep operating the farm. Our milk customers were referred to other local dairies. For a short time, the milk was sold to other local dairies, until arrangements could be made to sell the cows at auction.

While this picture is not very clear, it holds a special memory for me. One day while driving the old pickup truck over this bridge,

Figure 3 - The bridge across the first brook, in the back half of the night pasture.

the person riding with me saw a spider on his shirt. When I looked over to try to help remove it, I lost control of the truck and one wheel dropped off the edge of the bridge. It required a tow from the H Tractor to get it back on all four wheels. Even to this day, no one else believes the spider story.

This bridge also holds a memory for my sister Dot, as she had to have Uncle Ike's help when our horse Pete's hoof got stuck in a gap between two of the bridge rocks. Dot rode the horse frequently, including for herding the cows.

Up until the late 1940s, we relied on coal for heat, in the farmhouse and the milk house. The coal was delivered to a bin just inside the milk house door, and into a bin in the basement of our house. Of course, this meant that the furnaces had to be fed and cleaned multiple times each day. Once oil systems were installed, our labor was greatly reduced.

Figure 4 - Uncle Bert Bray's painting of the barns.

In a family with much artistic talent, Uncle Bert Bray was a particularly good artist. This picture, another depiction of the farm buildings' layout, is one of my favorites. To the left is the hay barn and at the center is the horse barn and one of the horse-drawn wagons. Behind the horse barn is the milk truck shed and on the right side of it is the shavings shed. Not shown but at the very right of the picture is the milk house. This view was from the driveway between our back yard and the Kings' yard, right in front of Bowles barn. I assume the barn was named for Uncle Mac Bowles who lived in the house in front of the barn prior to the Kings living there.

Prior to the availability of mechanical refrigeration, the horse barn was the icehouse. Its double walls were filled with seaweed for insulation. Ice was gathered from local ponds, stored in the icehouse, and used to keep the milk cold.

Chapter 2
Spring

Figure 5 - The gate leading to the plain, the first field mowed each year.

At Springtime, each year for many of the past years, I am reminded of activities that we did on our dairy farm. Usually, something will prompt those memories, and this year, the first week of May, I noticed some heifers in a nearby pasture in Ashford, just as the grass was starting to grow.

In Spring on our farm, major activities included fencing, plowing, harrowing, fertilizing, and planting. I was a mature adult before I learned there was a sport called fencing and it was in no way related to the fencing of my youth.

Plowing, harrowing, and planting are done in the Spring, as soon as the ground is thawed and somewhat dry. Plowing results in the ground being flipped over about 6 inches deep. This flipping helps to fertilize the soil, and harrowing then prepares for re-seeding. The harrow would be towed across the field, its dozen or so sharp disks cutting up the furrows left by the plow. Harrowing usually required

Figure 6 – Our Plow. The circular disks would slice the ground to start the furrow.

two passes to get the field to be smooth enough for planting. Once the field was prepared, seeding was done. Needless to say, birds were everywhere during seeding.

Our 'big' tractor was a Farmall H model, and it remains in the family to this day. Of all the jobs the H was used for, plowing was the most demanding, requiring full power for hours at a time to pull the plow across the entire field. After each field was plowed, this same tractor was used for harrowing, but that task was much less demanding. The primary fields that were plowed were the plain, the Northern plain and the four acre lot.

We had approximately 80 animals, including 28 milking cows, 8 dry cows, and 44 young stock. Each cow goes 'dry' each year for about 3 months prior to calving. Shortly after calving, she is returned to the milk barn. The young stock was made up of the calves up to about 6 months of age, young heifers (females) from 6 months to 15 months, and older heifers from 15 months to 24 months.

During the summer, these three groups were pastured separately, with the calves being kept close to the barn, young heifers in a pasture on our property, and the older heifers at remote sites belonging to other farms. The young calves were kept close by so that they could be fed grain, hay, and, in some cases, milk. Each year, we would keep one or two male calves that would be used for breeding in their second year. For safety reasons, we did not keep older bulls but did keep steers which we processed for beef.

In each of the pastures, every inch of fence had to be checked and repaired each year. Of course, this had to be completed before the end of April to accommodate the arriving heifers. I learned at an early age that cows, especially the young ones (heifers), are not easy to contain. It seems to me that they spend their first few days in a pasture inspecting the fence, and then routinely try to escape during the night. I was reminded of those days recently when one of two animals pastured down the street from my home in Ashford escaped on its second day there. I came upon the escapee as I was driving to do some errands. While I went to locate the owner, the animal proceeded to the neighbor's yard to do some exploring. As was the case 60 years ago, the animal was returned to its enclosure without much fanfare.

Fencing was done as soon as weather allowed, starting with the pastures closest to the farm. Fence posts were checked to ensure they were still solid. Barbed wire, at least three strands separated by about 12 inches, was checked to ensure it was tight. Our fence posts were crafted from scrap telephone poles, cut to length, and split into four posts per length. Posts were pounded into the ground and then the wire stretched from pole to pole using a tool called a wire stretcher to get the wire tight. Once tight, staples were used to attach the wire to the post. In some cases, corner posts required extra support posts to keep them solid.

We have all heard the phrase about greener grass on the other side of the fence. Cows seem to agree with that notion. With them regularly pushing on the barbed wire, both it and the posts took a beating, so it was rare to go fencing and not have to perform many repairs.

At our farm, the 'back pasture' was approximately 12 acres, and a portion of it was wet due to a stream that ran through it and eventually emptied into Long Island Sound. For most of our pasture areas, we were able to drive our pickup truck along the fence, making it more convenient for getting the tools, wire, and posts where we needed them. However, in the back pasture, there were a few areas where the wetness meant that we couldn't get the truck close enough and had to carry in the mending supplies. It also meant the uniform of the day included boots.

We had several pastures for our animals. At our farm, there was a small area of a couple acres just behind the tractor shed. This was for the young heifers up to about 6 months old. The back pasture, extending almost to Great Neck School, was used for heifers up to about age 15 months. We also had another area of about five acres directly behind the cow barn referred to as the night pasture, as the milk cows would be left there during the Summer whenever they were not being milked.

Another pasture lay to the south of our hay barn and west of Bowles barn. During the Winter, the dry cows and some of the heifers were kept in the enclosed lower level of the hay barn. As the weather cooperated, we were able to let them out for several hours during the day.

Figure 7 - The Mansion at Harkness State Park.

We also had a Summer pasture at Harkness Memorial State Park. The land had been donated to the state by Mary Harkness in 1950.[1] She and her husband Edward, who lived in Hartford, were well known philanthropists and established the property as their Summer home. When she donated it to the state, part of the property was designated to be a state park and the rest as a summer camp.

One of the prominent buildings at Harkness is the Rogers House, a large white building easily seen from Great Neck Road and part of the Summer Camp section of the park. While the house is part of Harkness today, it was known to be part of the Hammond Estate years ago, and was built for the Hammonds' caretaker.[2]

After the dairy went out of business, most of our property was sold to Verkades Nursury and eventually to the state to become part of the State Park. In total, now, the park boasts about 350 acres.

My sister Mary's son Matthew and his wife Katie were married in this mansion. Dot's son Ben and his wife Mary were married there as were her daughter Laura and husband Mike. It is open to the public for such events.

Another pasture was part of the former Hammond Estate, previously known as Walnut Grove and owned by Gardiner Greene Hammond, a wealthy Bostonian. Per Wikipedia, it served as

Figure 8 - The Mansion at Hammonds.

---

1. From Wikipedia.
2. From *An Illustrated History of Waterford, Connecticut* by Robert Bachman.

their Summer home and still contains the fabulous mansion that highlights the site. The property, sold to the town in the early '60s, is currently occupied by the Eugene O'Neill Theater Center, and by Waterford Town Beach.

For a few years, we used about 4 acres at the former Gallagher Farm, at the north end of Dimmock Road. Yet another pasture was the Jacobs property, directly across Dimmock Road from our farm and extending east to Great Neck Road at its intersection with Lloyd Road. The property now has perhaps eight elegant homes on it. Finally, we used a pasture on Cross Road in Waterford, a site that used to be a chicken farm. One of the least enjoyable jobs we kids did was cleaning the chicken coops on that farm. Uncle Ike loaned us out for that task. While it wasn't part of a pasture, Ike also loaned us out to clean another chicken coop on Shore Road in Waterford.

The animals were moved to pasture the last Saturday of April. The obvious significance of doing this on Saturday was that the kids would be home from school and able to help with the moves. Moving the heifers to the back lot could be accomplished easily by herding them through the night pasture, onto the plain and then to that pasture. The moves to the Jacobs pasture could also be done by herding those animals down our driveway and across Dimmock Road into the pasture.

The rest of the remote moves required loading those animals onto the 1948 Chevrolet truck and delivering them. This would require multiple trips, since we could only fit four animals per trip, and often fewer. We would also move the 'dry' cows to the Harkness pasture at the same time or perhaps the next week. As I drive through Waterford these days, I recall fondly those pastures and the work it was to make those moves.

In the movies, it is common to see horses being loaded onto transportation trailers, seemingly to their delight. Cows and heifers do not seem to look forward to the truck ride, so frequently resisted getting on the truck. Uncle Ike had quite a few tricks for getting them to comply.

Two other activities that happened predominately in the Spring were picking up sticks in the fields and hunting woodchucks. Throughout the Fall and Winter, especially during foul weather, trees would lose branches and limbs. These needed to be removed prior to haying season. This was one of my earliest paying jobs on the farm, beginning perhaps when I was 6 or 7. I think I was paid $0.10 an hour for that. Hunting for woodchucks was also a revenue arrangement, but not by the hour. We got paid $0.25 per woodchuck. The importance of eliminating woodchucks was that their burrows included large holes in the ground that could result in a broken leg if a cow stepped there. I must say that our attempts to minimize the woodchuck population was not very successful, as there were more every year.

Figure 9 - Moving the young heifers to an on-site pasture

Chapter 3
Summer

On the last Saturday of May, Summer started for us as we began cutting the grass. The first fields were cut for grass silage, so cutting and collecting was done the same day, with no need for drying. The difference between silage and hay is that hay requires several days of drying before baling.

We had a Farmall C tractor with a cutter bar on the back of it. The cutter bar was about 6 feet long and contained about 24 knives, each one about 3 inches wide and shaped like a triangle. Uncle Ike would sharpen the knives every morning before going out to do the cutting and again after lunch if there was afternoon mowing to be done. As the tractor moved through the grass, the cutter bar moved side-to-side, slicing the blades of grass so that they fell behind the bar. Since the cutter bar ran to the right side of the tractor, as shown in the picture below, the first pass would result in cutting grass six feet from the outside of the field. Therefore, after cutting the rest of the field, the final pass would be to go backwards to cut the grass closest to the wall that was missed initially due to the mower configuration.

Figure 10 – Uncle Ike Mowing with the C Tractor.

After mowing, the grass was raked into rows, the same day for silage and the next day for hay. The side-delivery rake was towed across the field, creating rows where the mowing had left the grass flat. Because this rake delivered the grass to the side, we had to start on the second mowed row, raking it into the first row, which meant that the first row to be picked up would be twice the size of all the rest.

Cutter bar mowing is rare these days, having been largely replaced by mowers that work more like rotary lawnmowers. They not only cut the grass but also leave it somewhat fluffed up to enhance the drying process.

Figure 11 - Side-Delivery Rake.

The rake shown in Figure 11, is like ours. The tow point is between the two large wheels and the gear arrangement is shown at the left side of the picture. The gears allowed the rake arms to spin faster than the wheels turned, and in the opposite direction when necessary. As the rake was towed, the arms, shown across the top of the picture and at a 45-degree angle with respect to the front wheels, would spin toward the front. The hay would be pushed by the tines on the arms and discharged at a point to the right side of the picture, resulting in rows.

We had two home-made tractors we referred to as mud buggies that were used to tow the rakes. One was created from a model A Ford, primarily driven by Don, and the other from a 1947 Plymouth car, primarily driven by Dick. Both had been built by Dad. As the rest of us got older, we took those over from Don and Dick. An interesting trait of both was that they had two transmissions, resulting in more power. If both transmissions were in reverse, the mud buggy would go forward. Another interesting trait was three speeds in reverse.

Figure 12 - Hay Loader.

We had two side-delivery rakes, hence the need for the two mud buggies. One of these rakes had the ability to turn in reverse, a trait that enabled it to be used for tedding to enhance hay drying. Tedding takes a smooth, raked row and fluffs it up so the air can more easily pass through it but then must be re-raked before baling.

Prior to 1954, we used a hay loader, like the one shown in Figure 12, to lift grass or hay onto a horse-pulled hay wagon, and then eventually onto one of two trucks. The hay loader had a series of tines, shown at the bottom of the picture, that picked the grass or hay up off the ground and lifted it to the bottom of its ramp. The hay would then be grabbed by tines on the bottom side of the 2x4's shown in the picture. There were six 2x4's attached and driven by a crankshaft that was operated by the main wheels. Each 2x4 would alternate up and down to pull the grass up to the top of the ramp where it would fall onto the truck. One of the crew would then move the grass or hay into a pile on the truck. This would continue until the truck was full. After 1954, we transitioned to a baler for hay and a field chopper for silage, eliminating the need for the hay loader.

Chapter 3A

Siloing

Each year, we would cut the fields at our farm, then Harkness and finally Hammonds. The first field to be cut at the Dimmock farm was the plain, a 10-acre field just behind the night pasture. It and the other fields at our farm were harvested for filling the two vertical silos and one square silo. The silos were filled with green grass, meaning the grass was mowed, raked, loaded to the truck, and blown up into the silo all in the same day.

Per Google, silage is grass or other green fodder compacted and stored in airtight conditions, typically in a silo, without first being dried, and used as animal feed during the winter. Many farms these days create corn silage, but we just created grass silage.

When the grass was initially mowed, it remained whatever length it was at mowing, typically about two feet long. Before it could be blown up into the silo, it had to be chopped into small pieces, perhaps an inch in length, since the blower could not blow full-length grass up the pipe. Prior to purchasing the field chopper, the trucks arrived at the blower with the grass at full length, as mowed. The blower paddle had large knives that chopped the grass into small pieces just before blowing it up the pipe. Once we put the field chopper into operation, it performed the tasks of chopping, filling each truck with chopped grass, thus enabling the permanent removal of the paddle wheel knives.

Once full, each truck was driven back to the silos and unloaded onto the blower. Its conveyor moved the grass into a rapidly spinning paddle that blew it up an eight-inch diameter pipe about 80 feet through a window opening in the roof of the silo. The grass then fell inside the silo to whatever level the previous filling had achieved. The paddle wheel and conveyor were driven by the H tractor, and, like plowing, another one of its significant challenges. The tractor had a power take off pulley that was connected to the blower paddle wheel by way of a leather belt. When the power take off was engaged, the pulley started spinning and, therefore the blower pulley.

Each silo had a three-foot diameter chute attached to the outside of it. Inside the chute, accessible from inside the barn, was a ladder for climbing to the top of the silo. Every three feet were doors that were closed off as the silo was filled. Each day in the Fall and Winter, silage was thrown down the chute. After each three feet of silage was removed, another door was exposed and could then be opened. Each door opened meant the climb to the silage level was easier. Our hired hand, Homer, was the one who mostly emptied the silos, climbing up all the way to the top, even as he got to be well into his sixties. Today's farmers benefit from automated unloaders that peel off the silage from the top and shoot it down the chute.

A constant battle during the silo-filling was preventing the pipe from getting blocked up. The silos were about 80 feet tall. If the grass was fed too slowly, there was a possibility of blockage, but sometimes if it was fed too quickly, it would block, too. Trial and

Figure 13 - Blower.

error were required each load to get it right.

This picture is a blower like ours. To the right side of the picture, between the rear wheels but not shown, is the pulley that was driven by the H tractor. Inside the circular enclosure at that same point was the paddle wheel and knives, spun rapidly by the pulley. This picture only shows one section of pipe, a gooseneck section, but to reach the top of the silos, additional sections of pipe, would be added.

The gooseneck section of pipe would redirect the stream of grass going up the pipe into the silo. That section would be at the top of about eight sections of straight pipe, each eight feet long. This 64-foot-long arrangement was assembled on the ground and then lifted into place using a rope and pulley. The pulley was permanently mounted near the window at the top of the silo and the rope was always in place and ready to lift the pipe when necessary. Once lifted, the gooseneck would be maneuvered into the window and the bottom straight section would be bolted to the blower discharge port.

At the center of the picture is the large tray that housed the conveyor that carried the grass to the paddle wheel. The bottom half of the conveyor can be readily seen at the bottom of the picture. A truck full of grass would be backed up to the tray and unloaded onto the conveyor at the bottom of the tray.

When we first started filling a silo, the grass could be leveled inside less frequently, as it naturally leveled as it fell. As the silo was closer to full, it would be leveled during each unloading. If it was not, the pipe discharge would get backed up, causing another pipe blockage.

Filling the square silo was easier in some ways, since the pipe was shorter and was at an angle rather than straight up. The square silo, however, required much more manual labor to level the grass while it was shot into the silo. In the round silos, the grass would disperse as it fell, leaving only a minor amount of work for leveling. The square silo was larger but not as tall. Being larger meant more labor for leveling.

I recall when I was about 8 years old, we purchased the field chopper and the Farmall F12 tractor, the tractor primarily to tow the field chopper. One unique feature of this tractor was that it had to be cranked by hand to start the engine.

The field chopper and baler replaced the hay loader, saving the labor previously needed to level the grass dropped by the hay loader onto the truck. The field chopper was towed along each of the rows created by the rake and would pick up the grass, chop it into small pieces and then shoot the grass into a truck traveling next to it.

Since the grass was chopped, the knives in the blower, as described above, were no longer necessary. After the truck was full, it would be driven back to the silos to be unloaded.

To contain the chopped grass, both the '42 and '48 trucks had side boards installed during silo season, as well as back doors. The trucks were also outfitted with a device to eliminate much of the manual labor previously associated with unloading un-chopped grass. A solid frame at the front of the truck body had cables attached to it that were pulled by a tractor to slowly unload the grass. Once the back door of the truck body was opened, a mud buggy pulled on the cables, dropping the grass onto the conveyor. Two of us would help pull grass from the truck using potato hooks, making it easier to level the grass in the conveyor.

My oldest brother Henry designed the unloading frame and tried quite a few versions before it was optimized, but all of them required less labor than unloading the un-chopped grass. Henry was always inventing new methods that we all benefited from.

Chapter 3B
Haying

Once the silos were full, we switched to haying, first at our farm and then at Harkness and finally at Hammonds. For use of the land and storage barn at Harkness, we were charged the exorbitant fee of $1 per year. Haying also meant removing the rear doors and the unloading frames from the two trucks. A full load of hay on the '48 was 112 bales while a fully loaded '42 was 96 bales.

Haying was different than siloing in one major way: a much longer view on the upcoming weather. When the forecast predicted three or more straight days of sunshine, we would mow one or more fields, depending on field size. The grass would begin drying immediately after cutting. Then, the next day, it was flipped using the side delivery rakes to continue the drying process. On the third day, we would flip it once more, and maybe even a third time, before baling it. If the weather cooperated, we could get hundreds of bales into the barn in a day, occasionally even more than a thousand.

Figure 14 - Raking with the H tractor.

While today's mowers and rakes are different from ours, it's easy to see the same activities as you drive by today's farms. Farmers have balers that shoot the hay bales into a trailer that is towed behind it. Other farmers have balers that produce round bales, weighing 700-1,000 pounds. Tractors are required to lift the bales, which are often wrapped in plastic and stored outside instead of in a barn.

The rakes are also different, typically using two or three pinwheels with tines on them to create the rows necessary for the balers.

Today's farmers certainly have weather to deal with too, just as we did. Their equipment is different, but they nonetheless rely on the sun to dry their hay. As I travel around, I still see their rakes at work preparing for the balers, whatever style they use.

Figure 15 - Haying with horses before the hay loader.

This picture shows what haying was like before the hay loader, baler and truck. Chubby and Prince, our two draft horses, are pulling the fully loaded hay wagon. It is easy to imagine the labor involved in throwing hay up onto the wagon, especially as the pile got higher. The horses also pulled the hay wagon when we kids wanted rides; I recall going on such hayrides with our horses pulling the hay wagon. Those going on hayrides these days probably have no idea of the origin of hayrides, or where the term came from.

By the early 1950s, the hay was loaded onto a truck by the hay loader, which was towed by a truck over the rows created by the side delivery rake. I recall steering the truck at an early age, perhaps 7 or 8. My legs were too short to reach the pedals so Henry would get the truck moving and rely on me to steer it over the rows.

Our hay barn had two large doors in front through which a fully loaded truck or hay wagon could be driven for unloading. When I first envisioned, as an adult, how the horse-drawn hay wagon traversed the unloading manoeuvre, I assumed that the horses would pull the load into the barn and then would be backed out of those same doors after unloading. A few days ago, my sister Dot told me that the barn used to have two large doors on the backside. This meant that the horses could pull the wagon into the barn and then pull the empty wagon out the back door.

I don't recall when it was done, probably in the early 50s, but those back doors were boarded up at some point and the ground behind it excavated so that a heifer barn could be attached to the back of the hay barn. Small doors were cut into the boarded-up large doors to make it easier to feed hay to the heifers from inside the barn.

Once the load of hay was positioned in the center of the barn, it was unloaded. Hanging from a rail in the peak of the barn, directly above the load, was a large hayfork (Figure 53, p.79), perhaps five feet wide when fully open. When it was open, it would be dropped onto the hay. Then the fork was closed around a large 'bite' of hay and raised up to the rail. The fork was lifted by a tractor during my days, but my brother Dick recalled that one of the horses used to lift the fork during his days. When the fork reached the rail, it would travel on the rail either to the south side of the barn or to the north, depending on how it was rigged that day. As the fork traveled along the rail, the trip rope was pulled when the fork reached the desired spot, dumping the hay down onto the upper floor of the barn, creating a hay mow or haystack. From that spot the hay would be thrown manually to the outer edges of the barn. Each fork-full was lifted into the barn until the truck or hay wagon was empty. Once the hay mows were full, we all loved jumping from one pile to another. Uncle Ike was none too pleased with such hijinks!

Once we started baling our hay, we had an elevator to lift the bales from the truck into the hay mow (upper part of hay barn), thus making the hayfork unnecessary. The bales were then stacked neatly in the hay mow. With the baler, we no longer needed the hay loader and hay wagon. I recall that both were sold to someone on Fishers Island and making that trip on the ferry. We also no longer needed the draft horses, so it's quite possible they also made that trip.

Figure 16 - Henry baling hay and carrying a rider sitting on the baling twine-distribution buckets.

As noted previously, we had two side delivery rakes, and both were used during the haying process. One had a feature that allowed 'tedding,' a process in which the reel turned backwards, flipping the hay up into the air to enhance drying. We would use the tedder whenever the hay got rained on. After tedding, the rake's gearbox would be reverted to normal and repeated the process so that the field would be returned to rows of hay and ready for the baler.

Around the same year that we got the field chopper for siloing, we also started baling the hay. Prior to that, as described above, we used the hay loader to load the dried hay onto the truck. Baling was more efficient in that more hay was packed into a smaller volume, making handling much easier.

For two years prior to baling our own hay, we had arranged with another farmer to bale our hay. Tiffany Farms, from Hamburg, CT, came to Waterford and baled our hay with their baler. For this, they got to keep every other bale. I think it was the Summers when I was 9 and 10 that I drove their truck while they picked up their bales. Leon Tiffany, Jr. was there most of the time, mostly with his dad. Leon's brother Jack, whom I met a few years ago at their farm, had some very fond memories of those days. One was that Mrs. Harkness used to host picnics each year in the late 40s for all the local farmers.

After those two years, we bought our own baler. The baler was pulled by the H tractor. It contained a power take off, a spinning drive shaft that drove the baler's plunger. As the baler was pulled over the rows of hay, the hay was scooped up and passed into the plunger chute. The plunger, with a stroke more than once per second, would pound the hay until a bale was the right size. At that point, and precisely timed, the needles would bring two strands of twine up around the end of the bale and the knotting mechanism would tie the knots and cut the twine. The baler then repeated the process, pushing the just-completed bale toward the discharge. Eventually, the bale would drop on the ground.

I mentioned "precisely timed" above because it was so critical: the two arms that carried the strings up around the end of the bales and up to the knotters had to quickly retract before the next stroke of the plunger, which would happen in less than a second. If the retraction didn't happen quickly enough, the plunger would collide with the arms, bending them out of shape. This collision didn't

Figure 17- Dad doing the baling, also with a rider sitting on the twine-distribution containers.

happen often, but when it did, it would likely require replacing those arms.

Quite often, we would not start picking up the bales until the whole field was baled, though this could change based on our other workload for that day. Typically, there would be two of us on the ground and one on the truck to pack the bales.

Unloading hay from the trucks was labor-intensive, but less so with bales. We had a conveyor that carried each bale from the truck up into the hay mow. From there, they were manually carried to where they would be stacked. Our cousin David was the best packer and was known to take down a full stack of bales, to start over if it was not straight enough. He went on to study Engineering at UConn and then completed a career as a teacher.

Another memory I have of Harkness is that the quality of hay was different depending on what area it came from. There used to be a golf course between the mansion and Long Island Sound. The hay from these fields was not good quality for feeding the cows, but was great for bedding them.

As mentioned above, after finishing harvesting the fields at our farm, we would move to Harkness, eventually to Hammonds, and

Figure 18 - The '42 almost full.

then back to our farm to begin the second cutting cycle.

Harkness has changed quite a bit since those days. One memory that comes to me whenever I go there is at the public entrance to the park, the western-most entrance from Great Neck Road. That entrance did not exist back then and the area on both sides of the current road made up the largest field that we worked anywhere. If the weather cooperated, it was common to bale and pick up 1,000 bales from that one field. That would be an exceptionally long day and would almost definitely result in a trip to the beach afterwards.

The beach we used was known as Common Beach. It was adjacent to the Bingham Beach, and it is at the very west end of what is now Waterford Town Beach. We went there often, especially after haying. There was no better way to get all the hay off us and refreshed. As a family, we have a right of way that will continue in perpetuity. The original right of way was for the collection of seaweed gathered for use as insulation in the icehouse. Since mechanical refrigeration was new back then and not commonly available, ice was gathered during the Winter and stored for the rest of the year. The notion that the ice would last all year is still inconceivable to me.

When the farm was purchased in 1847 by my great-great-grandfather Rufus Leverett Dimmock, it included a controversial seaweed right. Dot recently gave me a copy of an 1852 deed clarifying that the seaweed right was indeed part of the original purchase of the farmland.

Most of Harkness is mowed regularly now, instead of letting the grass grow for haying. Each time I see them mowing, I think about the haying we did there, trying to recall how many bales we might have gotten for each of those fields so many years ago.

There is one field at Harkness that brings back a specific memory. When I was about 14, Uncle Ike allowed me to do the mowing at Harkness, including driving the tractor on the public road from our farm to theirs. Uncle Ike reminded me that there was a woodchuck hole on the seventh pass around the field that directly lined up with the front wheels of the tractor. For the first couple years, I remembered his warning, but then, one year, I forgot. Sure enough, the front wheels dropped into the hole as Ike warned me. Getting out of the hole required complete removal of the mowing attachment to lighten the load.

A few years ago, I told this story to friends. One of them asked why we didn't just fill in the hole. He obviously didn't know how persistent woodchucks are.

We stored hay in each of the local barns. The barn at Harkness, which was located just east of the current west entrance, and the barn at our farm have both been dismantled. The Harkness barn

Figure 19 - Hay barn from the back side prior to its demolition.

was the biggest, holding well over 20,000 bales.

The barn at Hammonds was also used to store hay but is now an integral part of the Eugene O'Neil Theater Center. This barn had two large doors on the front and two at the back, meaning that we could drive a loaded truck in one way, unload the hay, and then drive the truck out the back side. We were able to store a few thousand bales in that barn. In the picture below (Figure 20, p.29), the barn looks to be in better condition now than it was then. I assume that it was renovated prior to its use in theater operations.

In my earliest memory, I recall that there were pigs kept in the small pasture at the lower level of the barn. I'm quite sure these were raised by the occupants of the farm at that time, but definitely not by us.

There are quite a few new buildings at both farms and a tennis court at Hammonds where we used to have our heifers pastured. One sad memory of that pasture is the loss of a prize heifer due

Figure 20 - The hay barn at Hammonds today, a prominent part of the Eugene O'Neill Theater Center.

to a lightning strike that knocked a tree down on top of her. That happened quite near the southwest corner of the current tennis court. The memory of her loss haunts me nearly every time that I drive on that part of Great Neck Road. I am sure that passengers in my cars are tired of hearing this story.

One of my earliest memories involves Uncle Ike taking me to mow a small field just off Niles Hill Road near its intersection with Lloyd Road. Each time I drive past there, I look down the driveway along the side of the house toward that field. What was special about it was that he used the two draft horses, Chubby and Prince, to pull the mower. The mower shown below (Figure 21, p.30) is not the one we used but is similar. My sister Dot has also recalled accompanying Ike to that same field to do plowing, with the horses pulling the plow instead. Ike also let her drive the horses, although just briefly. My brother Ed also remembers raking that field with a horse-drawn trip rake. After completing that task one time, the horse, Scout, took off and galloped back to our farm, perhaps a half mile, with Ed hanging on to the rake and horse for dear life.

This is a trip rake (Figure 22, p.31), like ours, sitting on a field near our house in Ashford. As noted above, it could be towed by a

Figure 21- Horse-drawn mower

horse, and ours had a seat for a rider. It was more often towed by a mud buggy or tractor. The trip rake was used to collect loose hay from a small field where the side delivery rakes didn't fit very well. In this picture, the tines are shown in the trip position, that is, up in the air. The rake would be towed across the field with the tines down touching the ground, gathering up the loose hay. It would be towed randomly across the field and then it would be 'tripped' to dump the collected hay. The process was repeated to amass more hay, dumping each collection to form a row. After all the field was collected, the row could then be easily picked up either by hand or with the baler.

Just to the left of the tow bar is a foot peddle. A rope was attached to it and would be pulled by the person riding the horse, or driving the tractor, to cause the rake to 'trip'. Our rake had a seat on it so a rider could sit on it and activate the trip using the foot pedal.

During my time on the farm, the trip rake was rarely used, but seeing one in the field brought back nice memories, as did Ed's and Dot's reminders to me of their experiences.

Figure 22 - Trip rake.

Another activity during the Summer was trimming along the walls and fences at Harkness and Hammonds, although not so much at our farm. The importance at the other farms was that they would be seen by the public, so Uncle Ike emphasized leaving the fields clean. After mowing, we had to make sure we went back to mow any spots that the mower missed on the first pass. After each field was finished with haying, Homer, our hired hand, would trim each area with a scythe. No one was as diligent as he at leaving the fields neatly trimmed, and no one was as proficient with a scythe.

Figure 23 - Henry and a helper throwing hay manually.

I'm not sure who Henry's helper is, but perhaps one of my siblings will know. I'm also not sure why they were handling the hay manually, because that was rare once we got the baler.

Figure 24 - Mom's painting of Dad driving the baler out to the plain.

Mom was good at many crafts, including painting. This is one that she painted showing Dad driving the baler out to the plain at our farm. This spot was just prior to the stream and to the right of the first hill. In the picture, it's easy to see how much wider the baler is than the H Tractor. This made it difficult to make it down a normal-width driveway. In the very spot that is shown, there were rocks on either side, making it necessary to slowly swerve around one to avoid hitting another.

In this picture, a bale is shown at the discharge chute. To the left and the right of the bale are two red boxes. These boxes held spools of baling twine and provided a seat for a rider or two, as shown in earlier pictures of baling.

Chapter 4
Fall

In September, once the haying wound down and the second cutting was stored in the barns, we started bringing all the heifers back from the pastures. Of course, if they calved during the Summer, they would be brought back immediately. We didn't always notice immediately when a new calf was born in the remote pastures. The new moms were good at hiding the calves; sometimes it took us as much as an hour to find them.

Fall was also the time to get all the animals back near the barn and get them situated for Winter. The cow barn could only fit 28 cows in it, so the rest had to be packed tightly into the bottom of the hay barn or held in an outside open shed at the back of the hay barn. There were two such sheds, one was a lean-to type shed attached to the hay barn. The other was a stand-alone shed in the night pasture. The dry cows and young heifers got priority for the inside spots in the hay barn basement. That meant the older heifers were left outside in the night pasture and could use the two sheds for cover.

Another major Fall activity was to bring all the farm equipment back for repairs and Winter storage. It was amazing to see most of it squeezed into the available tractor sheds. Most of those sheds are gone now, long since removed, but the last one to be built, and the biggest of them, is still standing today, and houses the H tractor that is still in use.

The C tractor also remains in the family. My cousin Donnie, who recently passed, used the tractor frequently and restored it to its as-new look.

Of course, Fall was also when all us kids went back to school, so our assistance to the farm efforts was restricted to afternoons and Saturdays. For any of us that participated in sports or other non-farm activities, farm assistance was further limited.

Occasionally, I would tell Mom that I didn't feel well enough to go to school. She'd say "OK, but you can't go out on the farm." Then

I'd change my mind and go to school.

Some of my school friends would come to visit the farm. These visits frequently happened during the Fall, just after school started. One of them recently reminded me that he made one of those visits.

Chapter 5
Winter

During the Winter, all our labor was focused on taking care of the animals: feeding them, cleaning their pens, and laying down bedding material. Feeding included grain, hay, and silage each day. Obviously, Winter also included cleaning up after snowstorms and patching and filling any holes in windows or walls. None of the barns was heated except by the body heat from the animals. Temperature control was accomplished by appropriate opening and closing of windows and doors, a significant challenge sometimes.

A major Winter task was to clean out the pens in the lower part of the hay barn, where many of the animals were kept during the Winter. We would change their bedding, hay mostly, several times a week. The cleaned-out material, including hay and manure, was dumped into a pile in the barn yard, and by the end of Winter, the pile was significant.

As Winter faded, and if weather cooperated, we would start to prepare for Spring. One of these preparatory activities was to use the C tractor with its front load bucket to move the pile in the barn yard into the manure spreader to spread, one load at a time, on all the fields at our farm. How was the weather pertinent, you might ask? We needed the fields still frozen, so the manure spreader, towed by the H tractor, would not tear up the ground. We tried to complete eliminating the pile before the ground thawed. Otherwise, there would be a several-week delay until the ground was dry enough.

Loading the manure spreader had some risks. Our tractors had their two front wheels close to each other with only a few inches between them. That configuration resulted in a truly short turning radius, but also in an increased risk of tipping over. The bucket was in front of the tractor. As it was lifted and the tractor maneuvered for dumping, the tractor could tip over, especially since the ground near the manure pile was somewhat rough. This happened at least once that I know of. Fortunately, no one was hurt, and the tractor was easy to get back on all fours again.

For many years, we got surplus shavings from Nassetta Brothers, a woodworking shop in New London. We used the shavings for bedding under the cows. In front of the hay barn was a garage for the milk truck and the knife sharpener. In front of that garage was a small shed where the shavings were stored. I don't recall exactly when we stopped using the shavings, but know it was when I was 10 or 11 years old. I know Nassetta's went out of business at some point, and perhaps that was the reason the shavings were no longer available.

Just like in the Fall, availability of the Dimmock kids to help on the farm was limited to afternoons and Saturdays. We didn't have many snow-induced school closings, but when they happened, we were happy to help with snow cleanup. Both tractors were used as well as traditional snow shovels. We did the yards around the farm buildings and then the local houses where our relatives and neighbors lived. After all the snow was cleared, we would frequently grab the sleds and carry them to the hill for some fun sledding down the back side. It was steep enough to get a surprisingly good ride. A great ride would be one that ended up on the ice on the small pond near the sand pit.

In Winter, the cows remained inside the barn unless the weather was good. I recall that they enjoyed going out after being in for an extended time, as evidenced by their jumping and otherwise frolicking. Their time outside was limited to just an hour or so during those infrequent excursions of the coldest part of the year. We used that time inside to clean the gutters and put down bedding. While these tasks had to be done every day, it was easier to do when the cows were outside.

My earliest recollection of cleaning the gutters is that Homer used a shovel, filling the wheelbarrow and then dumping each load into the manure spreader that was parked just outside the back doors of the barn. When I was about 12 years old, we installed an automatic cleaner, a conveyor chain that ran along the bottom of the gutter. Each day, it was turned on to clean the gutters, dumping the manure into the spreader, just as Homer had done manually before that. The spreader was parked outside the back of the barn and was then driven to the fields and emptied.

Another Winter task was delivering the hay that we sold. Each year, we had hundreds of excess bales, and customers that needed them. We would use the '48 truck to deliver as it had commercial plates while the '42 only had farm plates. Occasionally, we sold a cow or two. The '48 would be used for those deliveries, too.

Chapter 6
Managing the Cows

Farms in Waterford back then were much smaller than today's farms. We had 28 cows that were milked twice each day. Today's farms have hundreds, or even thousands, of cows, and they are milked three or four times a day. Cows, it turns out, produce more milk if they are milked more often. Some farms now have kiosks to which cows can walk up and get milked whenever they choose.

Uncle Ike, and our hired hand, Homer, milked the cows in the morning at 6am and then again in the afternoon at 4pm. We had two automatic milking machines that each would collect milk from a pair of cows. Each pair would take about 10 minutes, then the milking machines would be emptied into pails and carried to the cooler. The milk was dumped into a container on top of the cooler that funneled the milk onto its refrigerant-cooled tubes. The milk was collected from the cooler into 40-quart cans that were then transported from the cow barn to the milk house and into its walk-

Figure 25 - The pickup truck transporting cans of milk to the milk house.

in refrigerator.

One of Uncle Ike's favorite pranks during milking was played on visitors to the barn. It was common for him to coax a newcomer to get a close look at an udder. When the unsuspecting prospect got close enough, Ike would squirt the person in the face with milk. It always got a laugh from the other visitors, not so much from the victim.

In the Fall and Spring, the cows were left out for two or more hours, again as weather allowed. One thing that fascinated us all was that the cows would return to their proper stall when they came back in. If they did not, a quick shout from Uncle Ike would get them back in line.

In the Summer, after the afternoon milking, the cows would be moved from the barn to the night pasture and then to a grazing field for 30-60 minutes. After that period of grazing, they were moved back to the night pasture for the rest of the evening. Most of us used the pickup truck to herd the cows to the grazing pasture, but my sister Dot would ride our horse Pete to drive them and then gather them back up after the grazing time was done. For the rest of us, the pickup truck worked fine.

Also during the Summer, the cows stayed in the night pasture all night and then were brought back in before morning milking. During the Summer, they were in the barn only for milking, a couple of hours in the morning and again in the afternoon.

Cows were fed grain, hay, and silage twice during the day. Hay and silage were reduced during the Spring and Fall, then eliminated during the Summer when grass was available in the pastures.

I recall that, early in my farming career, grain was delivered to the farm in 80-pound bags. Twice a day, a bag was dumped into a hopper, which was then wheeled to each stall and grain distributed in the proper amount for that cow. In a labor-saving effort in the mid-50s, we installed a large grain bin outside the barn next to the silos. It had a chute that could be opened inside the barn to allow filling the hopper, which would then be wheeled to each cow for their serving. The grain bin was replenished about once a month from a bulk delivery truck.

Silage was manually thrown down from the silo twice each day, usually by Homer. He would climb up the silo ladder to whatever level the silage was at and then throw enough down the chute for the day's needs. When the silo was first opened in the Fall, the upper layer of silage was removed and thrown away until the good

Figure 26 - Cows going out after milking

silage was uncovered. Once the 'crust' was removed, the good silage was thrown down. After enough silage collected at the bottom of the ladder, it was manually shoveled into a hopper that was then wheeled to each cow.

The cows were held in place by stanchions with the manger in front of them, a water bowl for each pair of cows that they could fill by themselves, also in the front, and gutters behind them. The grain, silage and hay were put into the manger and quickly devoured.

Grain was distributed to each cow with a small trowel-like scoop, an individualized amount specific to that cow. Silage was distributed into the mangers, equally for each cow. Hay was stored in the hay barn. Each day, sufficient bales were moved to the cow barn for that day's needs. After the twine was cut, the bales were broken up and distributed in the mangers. Prior to bales, the hay had to be handled as loose hay, requiring much more labor.

During the late Spring, Summer, and early Fall, the 28 milk cows were kept in the night pasture when not being milked, as shown in this in Figure 27, p.42. This pasture began just behind the cow barn but also extended back behind the first hill, across the stream, and to the plain. During warm weather, the cows would be in the night pasture for most of the day and night, only going into the barn for milking.

Figure 27 - Cows in the night pasture.

Figure 28 below shows Henry's oldest son David feeding the young heifers in their pasture, the one just behind the sheds. If you're wondering how I know it's David, it's the slacks. No one else wore such loud slacks. David and all four of his brothers worked on the farm, but after my time.

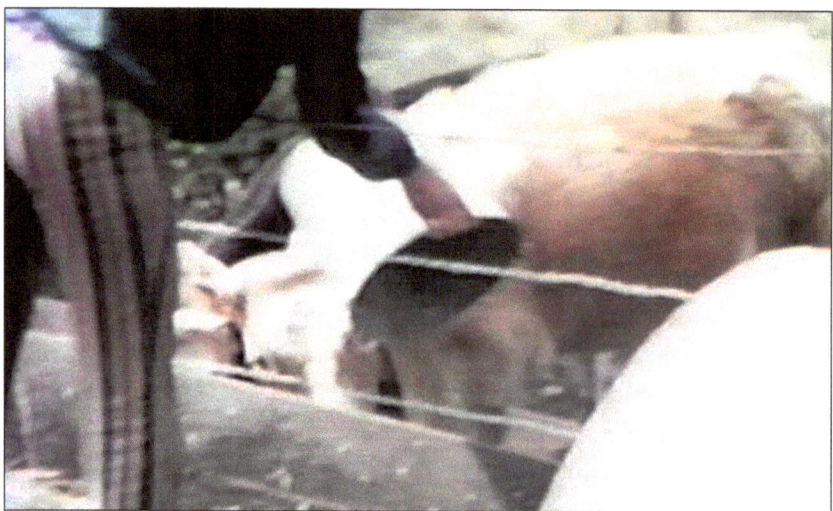

Figure 28 - David feeding the young heifers.

Chapter 7
Milk Processing

Milk was processed six days a week starting at about 9am. This processing was usually begun by Uncle Ike after his post-milking breakfast, while Dad was still on the milk route. When Dad returned, he would have his post-milk-route breakfast and then join Ike for the remainder of the milk processing. Whenever any of the kids were available, myself for one, we would take Ike's place in the Milk House so Ike could get on with haying activities for the day.

First, the empty bottles were washed in hot steamy water, steam-dried and placed upside down into the milk cases. Milk was then dumped from the 40-quart cans in the refrigerator into the bulk tank, enough to meet the needs for the next day's delivery. The milk was then heated up in the bulk tank as part of the pasteurization process. For pasteurized milk, the temperature was raised to 145 degrees for 30 minutes; the milk was then cooled and bottled. Around 1958, we started selling homogenized milk. For that, the temperature was raised to 150.

After the 30 minutes for pasteurization had elapsed, we pumped the milk through the homogenizer, over the cooler, and into the bottler. The bottler filled each quart bottle, and pressed a cap onto the top. From the same batch of milk, the pasteurized milk was bottled first, with the homogenizer in flow-through mode. After the pasteurized milk was all bottled, the homogenizer was changed from flow-through mode to pressure mode and pressure was raised to 1500 psi, thus homogenizing the milk for its bottling.

We also sold raw milk, and it was bottled after the pasteurized and homogenized. We had numerous customers that had a strong preference for raw milk. Some of these folks had well-tuned taste buds; they could detect the taste change in the milk in the Fall when we started feeding silage.

Figure 29 – The bottler, filling bottles and installing caps.

Two other products from the dairy were skimmed milk and heavy cream, both being big hits. The cream separator was used to create these products. It worked via a centrifugal force method, which used a six-inch stack of tightly coupled cup-shaped disks that spun at several hundred RPM. Milk was poured into the 25-quart container at the top of the separator. As the milk passed down into the spinning disks, skim milk came out of one discharge pipe and heavy cream from the other. The volume of skimmed was about 10 times as great as the volume of heavy cream. Once done, the skimmed milk was dumped into the bulk tank and pumped across the cooler and into the bottler, just as the other products. The heavy cream was manually bottled into half-pint bottles.

One feature of the heavy cream was how thick it was. If it sat in the refrigerator for more than a day, it could not be poured without the assistance of a spoon to get it started. Dad used to have heavy cream on his pancakes instead of syrup.

As the full, capped bottles came off the bottler, they were placed in milk cases, one dozen bottles to a case. Each case was loaded

Figure 30 - Milk stored in the refrigerator, 12 quarts per case, 18 cases per dolly.

onto a dolly until the dolly had 18 cases, at which point the dolly was moved into the refrigerator. Each of the full dollies had to be carefully stored since there also needed to be room for the 40-quart milk cans from the evening milking. The refrigerator was most packed on Saturday since we bottled enough for two days' deliveries, Sunday and Monday.

Chapter 8
The Milk Route

Each day, Dad loaded the truck with enough product for the day's deliveries, always bringing a little extra for good measure. Dad got up at 4:30am, started loading the truck at 5, left at 5:30 to start delivering, and did not finish until around 9am on good days, or as late as 10am some days.

His days were long, especially during the Summer, as he was active at the Goshen Fire Department, serving in leadership positions including as its first-ever President. It was common for him to get home after 10pm. Dad, Henry, Don, Dick and I would respond to a fire alarm whenever possible, based on what tasks we were working on. Some could not be interrupted.

Dad had two delivery routes that he alternated from one day to the next. That meant a customer would get their milk on Monday, Wednesday, Friday and Sunday one week and then Tuesday, Thursday and Saturday the next. One route ran primarily through Waterford and the other mostly through New London. I think we only had one customer in Niantic, Cousin Bob and his wife Peg. As we carried out the deliveries, we would stop periodically to rearrange the cases, moving the empties toward the back of the truck and the remaining full ones toward the front.

At some stops, there would be just one customer, handled by the runner, but at others, both Dad and the runner would have customers. Another of those lingering memories is one of those places where Dad dropped me, the runner, at one house then drove down one street, made a delivery, and then drove back up the next street to pick me up. Nothing memorable about that, except the day of a major snowstorm. As Dad was coming back to get me, the snow was so deep it was going right up over the hood of the truck. Because of the weight of the milk, I think, the truck never got stuck in the snow.

Most deliveries were straightforward, but at some stops, we had to deal with dogs and at others, we had special accommodations for the delivery. As I travel around Waterford and New London these days, more than 50 years later, I still recall a few of those customers and their special delivery requirements. Two that I remember had a small hatch door on the outside wall of the house. A box lay on the other side of the door. We opened the hatch door and put the milk inside the box. Inside the house was another door, opened by the customer, to retrieve their milk. While dogs were only an occasional problem, one home on Niles Hill Road did have a stealthy dog that would sneak up behind the delivery person and try to bite him. We were incredibly careful to avoid such an attack.

I was Dad's runner every other day for four years, from 7th grade through sophomore year in high school. At the beginning of my junior year, I obtained my driver's license. What a thrill it was for me to then be given the responsibility to take Dad's place driving the truck to carry out the Sunday deliveries. I think I did a good job at it except the one Sunday that followed one of my late-night Saturday dates. That morning, I started out ok but had trouble keeping my eyes open. Eventually, after the runner had to keep waking me, I had to return home and wake up Dad and ask him to cover for me. He did so and never brought that episode to

Figure 31 - Dad returning from the day's deliveries.

my attention again. From then on, I made sure not to stay out too late Saturday evenings.

Each day, when Dad finished deliveries, he had some breakfast and then went to the dairy to join Uncle Ike in processing the milk. That took until lunch time. After lunch, he would help with the day's farm activities except on Tuesdays. That was his day to travel to customers for collections. We left bills to people either weekly or monthly, their choice. Most customers would leave the money in the empty bottles the next day but some required home visits. These were conducted on Tuesday along with any banking chores, and it frequently took all afternoon.

One of Dad's enduring frustrations was how long his day was compared to 9-to-5 workers. He might see them on their way to work after he had been on his route for a couple hours, and then see them again in the afternoon on their way home while his day was still going on. This is a great example of the hard life of a farmer.

The milk route also had a few traditions—breakfast stops, for example. Two that I recall were coffee houses that are now restaurants, one currently occupied by The Yolk on Montauk Avenue and the other by The Recovery Room on Ocean Avenue. The stops were brief, but certainly long enough for a cup of coffee and a donut or muffin. We would frequently see the crews from Michael's Dairy and Radway's Dairy at these stops. There were fun conversations, sometimes resulting in more lengthy stops.

Another frequent stop took place outside the fence at the Coast Guard Academy when the cadets were practicing their marching. It was always impressive to see them and observe their improvement as the Summer gave way to Fall. This started in 1957 when our brother Ed was a fourth-class cadet (freshman), continued for his four years, and then started up again in 1964 when I was a fourth-class cadet. With "Dimmock's Dairy" painted all over the side of the truck, Dad's visits drew plenty of unwanted attention from the upper classmen inside the fence.

Dad was always good at smoking out our previous night's escapades. He might ask, "What were you doing in Stonington last night?" This happened to me quite a few times, and legend has it that it happened to my brothers, too. It seemed that he had sources in the other dairies, the police departments, and the fire departments, all of them more than happy to rat us out.

In the Introduction, I mentioned that lots of current events in my life remind me of my days on the farm. As I drive around Waterford and New London these days, I am reminded of areas that are so different now, Shaw's Cove area in New London, for one. Where there are now huge office buildings, then there were housing complexes. The area near the Coast Guard Academy is quite different, as is the area all along I-95.

I also recall some funny milk route stories. One of my predecessor runners told the story of an elderly customer who lived on the second level of a home at the south end of Jefferson Avenue, across from the CVS store currently on that corner. Each day when he went up the stairs to deliver her milk, she would ask him to come in to shut off the water faucet because her hands were sore. Each day, he would do as she asked, but began to wonder how the water got turned on in the first place. One day, he snuck up the stairs quietly just in time to hear her run to the faucet to turn it on before she hurried to the door to ask him to turn it off.

Another one of my favorite stories involves another of my predecessors. He just happened to be quite slow, and especially so in the morning. One disadvantage of the milk truck's panel configuration meant that it was difficult to see traffic if coming to an intersection where angled roads merged. Dad would ask if it was ok to make the merge. The runner would look back and say "yup (long pause), after the next car." Dad learned quickly not to go until after that long pause.

One customer on our New London route was a small grocery store on Willett's Avenue, just east of Montauk Ave. We would drop off a few quarts each delivery that he would then sell to his customers. We would also pick up any groceries that Mom had ordered. She knew we would be there by about 8:30 on alternate mornings, so she would call the store to place the order and the owner would have them all bagged for us by the time we got there.

One thing funny about this store is that the owner couldn't tell my voice from Carol's. When Carol called in the order, he would refer to her as Sonny, thinking it was me. She did not appreciate that.

I learned a lot from Dad during those mornings for the four years of my life when we delivered milk together. We talked a lot, but sometimes just carried out our respective duties quietly. As mentioned in the Introduction, I have a much better appreciation of those life lessons now than I did when I was living them.

Figure 32 - We gave a calendar to each customer every year.

The picture above shows a calendar that we were able to purchase for delivery to our customers each year. The cover was of course different every year. At the bottom of the picture, there was a place where we could indicate their delivery days. As mentioned previously, in the last few years of the dairy's operation we discontinued Sunday deliveries and delivered to one route of customers on Monday, Wednesday, and Friday and then the alternate route of customers on Tuesday, Thursday, and Saturday. Since this calendar had a spot for us to indicate which set of days customers' delivery would be, the pictured calendar would have been distributed in these last few years. Prior to about the mid-

Figure 33 - A sad note from one customer.

sixties, we delivered on Sundays too, meaning customers would receive milk every other day.

We didn't often receive sad notes like this one above. I was impressed with the author's optimism that his wife would be returning.

I did not recall how the milk route deliveries were transitioned until seeing this picture (Figure. 34, p.53). I don't recall Maple Shade Farms, but assume it was another of the local dairies, perhaps in Montville.

In the New London area, there were three companies that delivered milk directly to homes, namely ours, Michael's Dairy, and Radway's Dairy. We delivered milk from our own cows while Michael's and Radway's purchased their milk from the many farms in the area.

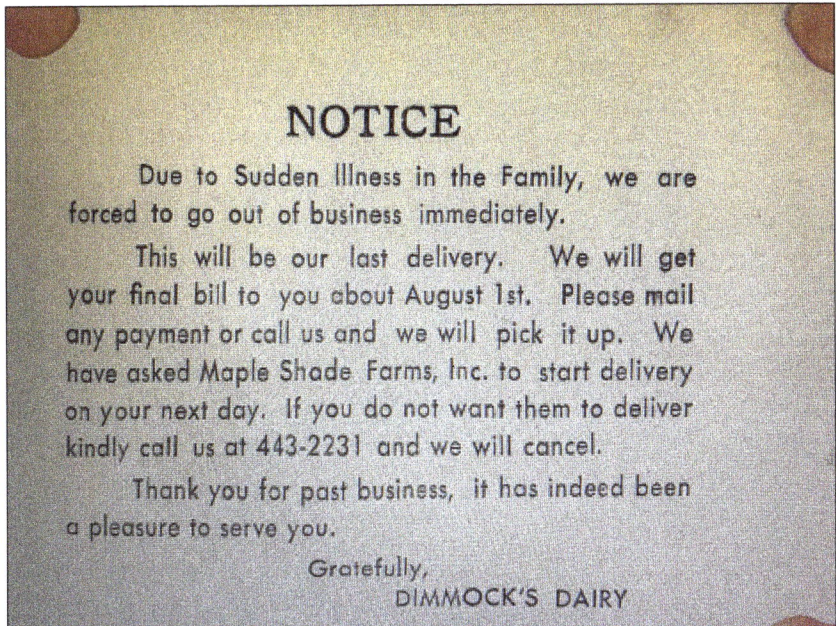

Figure 34 - This is the notice of business termination of Dimmock's Dairy.

Michael's Dairy is still open on Montauk Avenue at Mitchell College but only to sell ice cream. Their barn was recently reconstructed and maintained its historic look. The former site of Radway's Dairy is now the home of The Granite Group on Jefferson Avenue.

I've mentioned the runners that helped Dad, me for one, brother Ed too. He also hired local kids to help and would adjust the route to optimize the best time to pick them up.

I'll mention those I remember with apologies to those I can't. One was Doug? Rogers, another was Billy Starr. During my time, Ken Christian helped. Brother Al reminded me that John Merrill also helped.

Chapter 9
Maintaining the Farmhouse

The success of our farm was not only the result of the hard work performed as described in the preceding chapters, but also the hard work done by the ladies of the family. Our mom ran the household, raising with Dad eight kids, including five boys and three girls. The boys all worked on the farm, but the girls did not, except occasionally. Late in her life, Mom used to tell a story about our sister Carol. As Mom told it, Carol was standing outside the milk house door, which was just outside our kitchen window, stomping her feet on the ground and saying, "It isn't fair. The boys get to work on the farm and the girls have to help our mother!"

There's no question in my mind that the success of the farm would not have been possible without the contributions of Mom and our sisters. As I think back on those days, the girls helped clean the house, do the laundry and ironing, take care of the younger kids, prepare meals, and do the canning of fruits and vegetables as the gardens were harvested. Mom also had nice flower gardens all around the house. We had a small vegetable garden just outside the kitchen window and a much larger one behind the haybarn. Of course, there was no shortage of manure for fertilizing. We grew tomatoes, corn, cucumbers, string beans, radishes, and carrots. All of these were good, but especially the carrots. They were huge, and I have never seen carrots that big since.

When the carrots were harvested, we would spread them out on the lawn just outside the kitchen window and spray them with water from a hose to clean the dirt off them. After they dried, we put them into plastic bags and stored them in the extra refrigerator that we had in our basement. They would last for the rest of the year until the next Summer.

Another favorite in the garden was rhubarb, and the resulting pies were the best.

Mom and the girls also canned some of the vegetables and

made jam and jelly. We had crabapple and pear trees, blueberries, strawberries, and wineberries. All these contributed to tasty meals and desserts throughout the year. Homemade bread was another favorite, especially when covered by the homemade butter.

We made butter from the heavy cream produced in the dairy. We had a butter churn and used it frequently. My sister Carol told me that Mary, our youngest sister, used to make butter often. That prompted my memory of our brother Henry and his experiments at making things better. I mentioned above that he was always looking for ways to improve things. We used to hand turn the handle on the butter churn, which started out easy but got more difficult as the process went forward. Henry's idea was to use an electric drill to simplify the process. It worked well for a few minutes, but then things went awry, and we had cream and butter all over the kitchen. Back to the manual method.

As mentioned previously, we boys all took on responsibilities early, and the same was true for the girls. Occasionally, my sister Dot was left in charge of the younger sibs, including me. Shortly after she had received her driver's license, and while Mom and Dad were away on vacation, Dot asked me to get a few cucumbers from the garden. This required a knife. I know how the accident happened, but no one believes my version of the story. Suffice it to say Dot's driving skills were demonstrated very well as she rapidly drove me to the hospital emergency room to have the knife removed from my neck.

Mom was a great cook, and we all looked forward to Saturday or Sunday evening dinner. Our meals frequently included beef because we raised two steers each year for that purpose. Each April and October, we prepared one of them and arranged with a local butcher to cut and package the meats. We would split up all the packages between our family, Ike's Family and our extended family. Our share was stored in a freezer in our basement, right next to the carrot-filled refrigerator.

As mentioned, at our weekend dinners, we would often have steak. These meals frequently included guests, and it seemed like the table always had room for more. Our brother Ed graduated from the US Coast Guard Academy. During the four years he was there, he would bring some of his classmates to the farm during their liberty hours. Frequently, they would be there even if Ed wasn't. On Saturday evenings, there would be no rush to get back since their liberty ended at 1am Sunday morning. On Sunday evening, though, they had to be back by 7pm. This meant that dinner was earlier than usual, and always resulted in a mad scramble to get everyone into the car by 6:15 so they wouldn't be late. Dad would drop them off at the North Gate and watch them sprinting down to Chase Hall to beat the deadline to avoid demerits.

The cadets were there at the farm to do a little relaxing from the rigors of CGA life. Sometimes, they would go hunting for small game. The meal one night was to be fried chicken. Mom exclaimed that there were a couple funny-looking chicken parts in the pan. It turned out to be squirrel parts brought back from the day's hunt.

Cadets were not allowed to have cars until their senior year, so the farm turned out to be perfect for hiding those cars that were procured a little ahead of schedule. A story I heard recently was that Dot had to go to the Academy late one Saturday evening to move one of the cars that had been stashed on an adjacent street for the few hours between liberty end Saturday night and liberty start Sunday morning. Word had gotten out that there would be some extra police surveillance on those streets looking for those illegal cars.

As mentioned, Mom had great flower gardens. My son Keith recalls visiting the farm and checking out Grandma's flowers. She often gave him a flower or plant to take home with him.

Our niece Laura and her husband Mike bought the farm after Mom passed in 2003. They have been great at keeping the house and grounds the way it was while we lived there. She too has marvelous gardens and has made significant improvements to the house and the area behind the barn. My sister Dot bought Homer's house when he moved out. Her husband Clay always has a long list of projects that Laura and Dot have him signed up for.

When we were kids, we had a Fourth of July picnic that included Mom's side of the family, her mother and sisters and their families. It was common to have 40-50 people there and included card games like pinocle, setback, or hearts, as well as walks to the beach. Pinocle was the preferred game until Dot's first husband Ben introduced them to hearts and setback in the late '50s. We are so fortunate that Laura and Mike still host these picnics each year.

Each year, Mom would take us kids on a weeklong camping vacation. (Dad couldn't go with us due to the milk route.) One Summer, Carol and Mary developed whooping cough. This meant we couldn't go on our usual vacation. Mom set up a campground for us in the back lots and it worked out to be a great alternate site. Just like it was at home, the girls helped Mom make these vacations very enjoyable.

One of Mom's practices that still amazes me is how she coordinated the laundry for all 10 of our family members. Each of us had a color assigned to us. Mom sewed an X in our clothing in that color, each undershirt, each pair of underwear and each pair of socks. She also assigned us towels and face cloths of the same color. It was labor-intensive to implement, but highly effective.

Mom was so important to all of us, but at her wake, I heard two stories that exemplified how much she meant to members of our extended family and others, too. First was by our cousin David who told us that Mom was instrumental in getting him interested in theater. When he was high school age, Mom took him to his first play. He was amazed that she would do that, even though she had 8 kids of her own.

The other story was one told by Retired Coast Guard Captain Joe Anderson, one of Ed's CGA classmates. He said that he wouldn't have made it through the Academy if it hadn't been for having the opportunity to spend time at the farm. He was from a small town in Texas and was homesick until he started visiting the farm during his liberty time. The Andersons remain friends of our family to this day. Mom and Dad developed life-long friendships with the cadets and their parents. I became interested in attending the Academy after hearing so much about it from Joe and many of Ed's other classmates.

Throughout our younger lives, Mom demonstrated her talents as a seamstress and a gardener, and of course, as a cook. As the kids got older and left home, she began painting, including several of the pictures in this book. Her paintings became presents to her children, nieces, and nephews. Some of them also included frames made by Dad.

As stated earlier, Mom and her daughters, our sisters, were integral to the success of our farm and our family.

Figure 35 - Mom working in one of her gardens.

Chapter 10
Fun on the Farm

A. Snow Sledding

Figure 36 - Prince & Chubby pulling the sled for a ride in the snow.

I do not recall riding this sled, but this picture shows some of my older siblings enjoying the snow this way. The extra weight also helped the plow to be more effective. This also provided a good workout for the horses. In the background, upper left is our house, prior to closing in the porch, which was accomplished in 1955. On the right side is the corner of the milk house and in the background to the left of the milk house is the King house. The Kings' children, George, Roger, Gene, and sister Audrey, also worked on the farm. Gene worked during my time on the farm; the others all worked prior to my time.

The King's house had previously been occupied by our uncle Mac Bowles, brother of our uncle Jack Bowles, who was the husband of Mom's sister Gert. When Mac moved to Cape Cod, the Kings moved in. Years later, our niece Laura and her family lived there. (They now live in the farmhouse.)

To the south of the King house, though not visible in this picture, was where my cousins Shirley, David, and Wayne lived. They also helped on the farm. Shirley and David were the same age as Ed and Dot, and Wayne was a year younger than me, the same age as Carol. Wayne and I worked together on the farm, and were best friends when not working.

B. Henry's House and 4H Club

Henry's family lived in a house at the end of our driveway. Henry built the house, and it's still there, although it's been expanded by the new owners to more than twice its original size. Since Henry's lot was usually wet prior to the house being built, he brought many loads of fill from the back Lots to raise the ground level. We had a dump truck back then, and he used the bucket on the C-Tractor to load the truck. Once again, I was trusted to drive the dump truck back and forth from the back lots, delivering each load of sand to his yard while he stayed behind to prepare the next load.

Henry was incredibly talented and inventive. In addition to building the house, he did all the electric wiring and plumbing in the house. After it was built, he retrofitted it with double-side, back-to-back fireplaces, one in the living room and one in the kitchen. He also was ahead of his time in building a heat recovery unit into the chimney to help heat the basement.

Uncle Ike was our 4-H club Leader. Our group had about 10 members. 4-H clubs were popular then and still are, comprising folks who are interested in farming activities. 4-H stands for Head, Heart, Hands, and Health and the organization's goal, per Wikipedia, is to develop citizenship, leadership, responsibility, and life skills in youth. They claim a membership of 6.5 million, ages 5-21.

Competitions were held annually, Including the raising of farm animals and food products such as fruit and vegetables. Winners of the local competitions were eligible to compete at the Eastern States Exposition in Springfield, MA. I recall that Dick's heifer won the Grand Champion award once at the local level and went on to Springfield.

There were horse or oxen pulling contests at the 4-H fairs. The competition would be to see which team could pull the most weight on a sled. I learned recently from my brother Ed that Uncle Ike was

Figure 37 - Ray, Carol, John Starr and Mike Guilfoyle getting ready for the 4-H Fair; Uncle Ike in background.

renowned for his handling of his horse pulling teams, winning at every level while treating the horses well.

Each year, we would prepare heifers and show them at the annual fair in North Stonington. As part of the rules, we had to 'raise' the animals as well as prepare and train them. To show well, we had to train them to stop with their front feet next to each other, and their back feet, too. Judging was based on how they looked and how they behaved. To be able to say we raised the animals, we had to pay Ike for their grain and hay. He didn't charge us very much, but wanted to make a point about ownership involvement.

Our co-leader was Dolores Loprinze. As a girl, she had been a farmer on the Anderson farm, which was just to the west of the entrance to Harkness, right across from the big red house, then owned by the Buell family. Dolores' kids, Daneen and Owen, were also 4-H members in our club.

Mom was prolific in saving information about family and friends. One document I saw recently was a story written by Dolores in 1963 for The New London Day newspaper's 25 years ago segment. During the '38 hurricane, there were massive power outages. With no power, the cows had to be milked by hand, so Dolores was commissioned by her dad, Mr. Len Anderson. Days later, after the power came back on, he convinced her that she had to keep milking them by hand since they had gotten used to it. She fell for the ruse for quite a few days after power had been restored.

Mom also saved school items such as pictures, report cards, and memos from the principal. One of those was embarrassing for me when I saw it recently, as it reflected incidents of my bad behavior at Clark Lane Middle School. From my mom's example of saving, I kept information for my kids and their school history. A few years ago, I brought my son Keith's folder to him when I visited him in VA. When he saw it, he begged me not to show it to his sons.

C. Riding Pete

Figure 38 - Ed, Dick, Dot and another child riding Pete.

I think this picture is of three of our oldest four siblings, and one other kid, riding on our riding horse, Pete. I think the one to the left is Ed, the third from the left is Dot. One of the others is probably Dick but I'm not sure who the fourth kid is. I am guessing, based on their ages and the car in the background, that this picture was taken in the late '40s.

D. Power Snow Sledding

Figure 39 - Kids going sledding the easy way.

This became a popular way to go sledding after the draft horses had been sold. This is the H Tractor towing a whole line of sleds. It's not possible to see who the kids are, but they very likely include some of Dot's kids and perhaps a few of Henry's. I think this picture was taken, at the end of our driveway, with Henry's car in his driveway, Larry Cochran's house in the top center, and Denny Haskell's house in the top right. Larry and Denny were regular helpers on the farm. Larry's wife was Kathy Tynan Cochran, and her brothers were runners on the milk route.

E. Chickens

Up until about 1951, we had chickens and roosters. I think this picture is Ed. I'm not sure who the smaller boy is, but it's probably Dick. Ed looks to be 8 or 9 years old, meaning the picture would be from the mid-'40s. If the younger boy is Dick, then the picture is from around 1945.

My lingering memory of the chickens is that I got attacked by a rooster when I went into the chicken coop. I think that happened in the early '50s, when I was 4 or 5. While I don't recall this, Dot and Ed have both remembered that the chicken coops caught on fire and burned to the ground. At the time, our gas tanks were right

Figure 42 - Ed and Dick feeding the chickens.

nearby and were in danger of making the fire even worse. I'm quite sure that the fire put an end to our raising chickens.

Another memory of chickens is when we harvested them for our dinners. That's when I learned of the phrase 'running around like chickens without their heads.'

F. Summer Evenings

During the Summer, the neighborhood kids would gather around the farm. Softball was one of our favorite activities. On the back side of the 'first hill,' there was an area large enough for a softball field. The infield area was somewhat flat even if the outfield wasn't. If we had enough kids, we would squeeze in as many innings as we could, starting around 7pm and ending before sunset, or when we just couldn't see anymore.

Two significant hazards on the softball field were rocks and… did I mention it was part of the night pasture???

Another favorite pastime was a game we called Kick the Can. It was a type of hide and seek with plenty of places to hide. The kid who was designated as "it" had to find each other kid. While the "it" kid was searching, others could come in and kick the can to free

Figure 41 – Dad giving a ride to Al's wife Judy and their daughters Christina and Andrea.

those who had already been found. We would play that game for hours, especially when the cousins came for an evening visit.

Our uncle Bill Perth would bring his daughters Jacki and Bobbi to the farm each Monday evening. His purpose was to pick up a week's worth of milk, but it also gave him time to visit with Mom and Dad, and equally important, a chance for the girls to join us kids for whatever evening game was on. Sometimes Jacki and Bobbi also brought two of their girl friends, the Deshefys. Our neighborhood boys would make it a point to be there on Monday evenings.

Figure 42 - Bowles Barn after collapse of its first floor.

Bowles barn lay to the south of the hay barn and was used for extra storage until the day in the early 1960s when the first floor collapsed. I don't recall what was stored in the barn, but it would likely have been hay, although this picture doesn't look like hay bales to me.

Audrey King lived next door to us. She had a horse and built a small barn for him and a small pasture just to the south of the Bowles barn.

Figure 43 - View of second hill seen from the back of the plain.

The plain was 10 acres in size and there was a smaller, 5-acre field to the right and a 4-acre field to the left. There is a fence post near the right edge of this picture, marking the fence between the plain and the 5-acre field. The 4-acre lot is not in the picture but is to the left of the hill.

To get to the camp site that Mom created, or to do fencing in the back pasture, we would go through this gate and to the right around to the other side of the hill. The driveway continued straight back toward the quarry and then a short driveway on the left took us to the camp. Also in that area was a cabin built by Don and Dick. A few years after building that cabin, they built another one to the left of the first hill, cutting down oak trees and taking them to a local sawmill to be cut up into boards for the cabin. I recall that building with oak required using a small sledgehammer to pound in the nails. Using a regular-sized hammer would frequently result in bent nails. They enjoyed the cabin for a couple years until they moved on at which point my friends and I took it over. After we moved on, Al and his friends occupied it.

There are plenty of stories to tell about the extracurricular activities that were part of our overnight stays at the cabin. I know quite a few from the years my friends and I stayed there. I'm sure there are as many from Don & Dick and an equal number from Al & his friends.

Figure 44 - View of the farm from just before the second hill.

This view gives good perspective on how big the farm was. Directly in front of the gate and proceeding toward the silos is the driveway between the 10-acre plain and the 5-acre northern plain. Not shown but to the right of this picture was the Anderson farm. In my time, we used their fields for pastures. In the Summer, we would also graze the cows there for 30-45 minutes in the evening prior to moving them back to the night pasture.

A stone wall is prevalent in this picture, but there were many of them throughout the farm. When Verkades bought our farm, they buried all the stones so they could have larger open fields for their nursery.

Another memory from this picture is the races we would have in an old car. We would start at the other end of the plain and take off from that gate to see how fast the car would go before we slammed on the brakes to avoid running into this gate.

Figure 45 - Mom's Painting of Common Beach.

After we kids were grown and the dairy went out of business, Mom took up painting. This picture shows the area of Harkness known as Common Beach, a small area at the west end of the Waterford Town Beach. Crossing over these rocks takes you to the restricted portion of Harkness, known as Camp Harkness, dedicated to citizens with disabilities. The rocks were fun to climb and a good place to go crabbing. Not shown in the painting, but perhaps 20 yards offshore and to the left of the area shown, was a 'big rock'. At low tide, we could walk almost all the way to the rock, but not at high tide. Legend has it that the rock also has a small underwater cave on its south side, the side facing Fishers Island. I know it has a small indent, but not sure about a cave.

Mom was a stickler about water safety. When she started taking the older kids to the beach, she decided she needed to get her lifesaving badge. I don't know that she ever had to use that skill, but I do know that she required us to demonstrate that we could swim before allowing us to go out to the 'big rock'.

Mom also learned how to drive the '48 truck so that she could transport as many kids as wanted to go to the beach. We routinely had 40-50 relatives at our July 4th picnic, and frequently 8-10 of them would take Mom up on her offer to drive them.

Figure 46 - Common Beach - Recent photo to compare with Mom's painting.

Figure 47 - Common Beach – Recent photo showing 'Big Rock' on left side.

Figure 48 - The creek between Ocean Beach and Waterford Town Beach.

In between Ocean Beach and the Waterford Town Beach is a creek. During the Summer, we visited this creek often, even crossing over to Ocean Beach from time to time. We also watched fireworks from the point, a bluff just behind where this picture was taken. One significant memory from this spot involves one day in about 1952, when I was around 6 years old. A bunch of us kids were at the creek along with a few of the older siblings. That day, we decided to cross. It turns out the creek is deeper, even near low tide, than a six-year-old is tall. Next thing someone noticed, only my hands were visible above the water. It turns out, not only is it deeper than a six-year-old, but the current is rapid at mid-tide as this creek fills and empties into Alewife Cove, a relatively large body of water. Perhaps we wouldn't have had this episode if the sign below had been posted then instead of 70 years later.

Figure 49 - Public Notice Sign between Ocean Beach and Waterford Town Beach.

G. Lloyd Road Sports

When I was in junior high school, I started playing football with the neighborhood kids. We started out playing on Wayne's front yard, but while it was flat, it was a small field. Then we discovered a field on Lloyd Road, and quite a few more potential players. It even got serious enough that I bought a helmet and shoulder pads. (More than once, I've been accused of playing too many games without a helmet.)

What reminded me of these games occurred a few years ago at a Waterford High School reunion. I didn't attend that school, but my wife did. One of the men at our table was Malcolm Williams. We started sharing memories from our youth and he reminded me that we used to play football in his backyard as well as in the yard adjacent. The yard wasn't as flat as Wayne's yard but together were much larger.

Figure 50 - Wineberry at Bluff Point.

Another memory of Lloyd Road that we talked about was the prevalence of wine berry bushes. He said his wife didn't know what they were and didn't believe they existed. A few days after the reunion, while walking at Bluff Point, I spotted this one wineberry and sent the picture to him so he could show her what they look like. We had many wineberries at our farm too, and jelly throughout the year.

H. Motordrome

Behind the first hill, there was a bowl-shaped area approximately 35 yards in diameter. This bowl resulted from the area previously being a sand pit. Over the years, even going back to when Dad was a youngster, they used to drive a truck around the inside of the bowl to see how fast they could go without losing someone overboard. Again, over the years, the rides ended with fewer riders than it started with, not just in our days, but also in Dad's and Ike's days.

I. Horse Chestnut Battles

Just outside our kitchen window was a large horse chestnut tree. Toward the end of the Summer, the tree would drop its fruit to the ground, and we would gather it up as quickly as possible for the inevitable battle with Dick and Donnie. For those who don't know, horse chestnuts are about the size of a golf ball, perhaps a little bigger, and have small spines on the outside of the husk. When they started dropping, it was a race to see which team could gather up enough ammunition.

One year, when the chestnuts started falling, we were surprised that our opponents weren't going after any, so we thought the battle would be a landslide in our favor. What we discovered, obviously too late, was that they had discovered several trees at Harkness, and gathered way more ammunition than we could from just the one tree. It was a landslide all right, but not in our favor.

J. The Secret Room

During the Winter, we would hang out in the basement of the hay barn from time to time. From inside the barn, there was a passageway to the square silo, as that is where the silage was piled up for the dry cows that were kept in that barn during that season. Over the years, we noticed that the north wall of that passageway seemed to be deteriorating a bit, as the stones seemed loose.

One day, we bumped up against one of the stones and it fell, not outside into the passageway, but down the other way, to the north. It finally occurred to us that it had fallen into a room, one that we had no idea was there. Prior to our discovery, I had never heard of it from Dad and Ike, nor from the older siblings. What I recall the explanation to be was that it was part of the Underground Railroad. To this day, I'm not sure whether that's fact or fiction, but it's certainly a reasonable explanation. The room was relatively large, perhaps 10 feet square, with no obvious access to it.

K. Sunday Rides

Dad worked 13 days in a row, with alternate Sundays off. On that alternate Sunday, we would frequently go for a ride in the car. It might be a trip to Scott's Orchards, or perhaps a visit to our cousins. Occasionally, these rides would also include a picnic at a roadside picnic area. One of these areas is still in existence today, on Route 85 in Salem, near Burnett's Country Garden. When I travel that road, I am reminded that Mom would prepare for the ride by making sandwiches, drinks, and dessert for all of us.

As mentioned above, there were 8 kids in our family, so I am sure that we all couldn't fit in the car at the same time. While my memory is not clear about which of my siblings stayed at home, I know that the car was full enough that I rode on the spot above the rear seats, the shelf right in front of the rear window. I must have been only 3 or 4 years old, and certainly didn't have a seat belt on, not only because of where I was but also because cars had no seat belts back then. Now that I've pinpointed my estimated age for these rides, it would have been before our two youngest siblings were born, so there were only 6 kids, but still quite a crowd.

L. Boy Scouts

Dad was a Boy Scout leader as a young father, but that was before my time. I recall that Dick was in the Boy Scouts and advanced to Star Scout. I was also in the Boy Scouts, a member of Troop 29 in Waterford, but only advanced to First Class Scout. Two of Henry's boys made it all the way to Eagle Scout.

We used to go on camporees from time to time. Two of those are memorable. On the first of these, we were at Gardner's Lake and had to endure a blizzard that snowed us in for two extra days. We helped to clear the snow from the roads at the campsite.

The second is memorable even to this day. We camped at Miller's Pond in Waterford. This pond can easily be seen while driving on I-395. It's to the south side of the highway, a mile or two from I-395's beginning. As I drive there these days, I can still spot the exact location of an accident involving my friend Mike Guilfoyle. At this location, a short jump is required to get from one shore over the water to the adjacent shore. Mike made that jump but slipped and fell with his knee directly on a small rock. The cut required a trip to Lawrence and Memorial Hospital and seven stitches to close the wound.

Chapter 11
Memories Now and Then

Figure 51 - Farm Layout.

Shown above is a Google Earth view of the property where our farm was located. For perspective, I've labeled the Pond at the Quarry, Great Neck Road and Dimmock Road. I've also attempted to show the various landmarks mentioned within the book. Sizes and shapes are approximate but relative positions are accurate.

Figure 52 - View of the farm from the first hill.

This is a view of the farm as viewed from the first hill. This was taken after the business was shut down. It's included here because it shows some items mentioned elsewhere but not shown in pictures. The silos are shown, although it appears there is only one as the second one is directly behind the first. This picture gives good perspective of how large and tall they were.

Next to the silos is the grain bin mentioned earlier, almost as tall as the silos. Also shown are the square silo, and the two heifer barns, one attached to the hay barn and one a stand-alone unit. These provided protection from the weather, especially during the Winter. The attached shed also made it easier to feed hay as there was a door from the haybarn.

Figure 53 - Hay Fork.

This picture shows a hay fork in the closed position. In operation, it is opened and then dropped onto the hay on the back of the truck or hay wagon. Then, as it is lifted, it collects a large clump of hay. The fork is then lifted, either by a horse or by a tractor, off the truck and up to the trolley track that runs the length of the barn. The eye locks into the carriage that runs along the track, carrying the hay to the desired spot in the hayloft.

Figure 54 – Son Jeff and granddaughter Delaney sitting on the C Tractor.

In the photo above my son Jeff and granddaughter are riding on a C Tractor.

Cousin Don renovated the C Tractor quite a few years ago. He displayed it at Waterford Days each year. Based on estimating Delaney's age, I think this picture was taken in 2015. As you can see, the tractor looks like new.

Don owned a couple of draft horses, either Belgians or Clydesdales, over the years. He displayed them at the annual celebration, too. He said that kids used to love the big horses, and the horses were gentle around kids.

As mentioned above, Don's widow Ginny still has one horse and the C Tractor.

Figure 55 - A plaque on the wall at the Waterford Museum.

The bottles shown in the bottler picture on page 44 (Figure 29) are 'square' bottles as shown in Figure 55 above. In approximately 1956, we had to change from round bottles to square due to lack of sufficient supply of the round bottles. This required a major change to accommodate the new ones. The bottles were new, and the bottler, the milk cases, the caps and capper, and finally the milk case dollies were all new. Then the question became what to do with the old bottles? We ended up sending them to the dump. Each week, we would bring several dozen bottles to the end of the driveway to be picked up by the garbage truck. This went on for quite a few weeks. We had no idea that each of these bottles would come to demand perhaps $5 at antique shops 50 years later.

Figure 54 - Family picture at a recent 4th of July picnic, probably 2019.

Each year, during the July 4th celebration, we gather in front of the cow barn for a picture; I think this one was taken in 2019. Nephew Ben is standing to the far left. Behind him is one of niece Laura's gardens. It sits where the hay barn used to be. As the nieces and nephews become grandparents, it becomes much more difficult to keep track of everyone.

Dot's daughter Theresa and Dick's wife Diane accept the overwhelming challenge of orchestrating the picture each year.

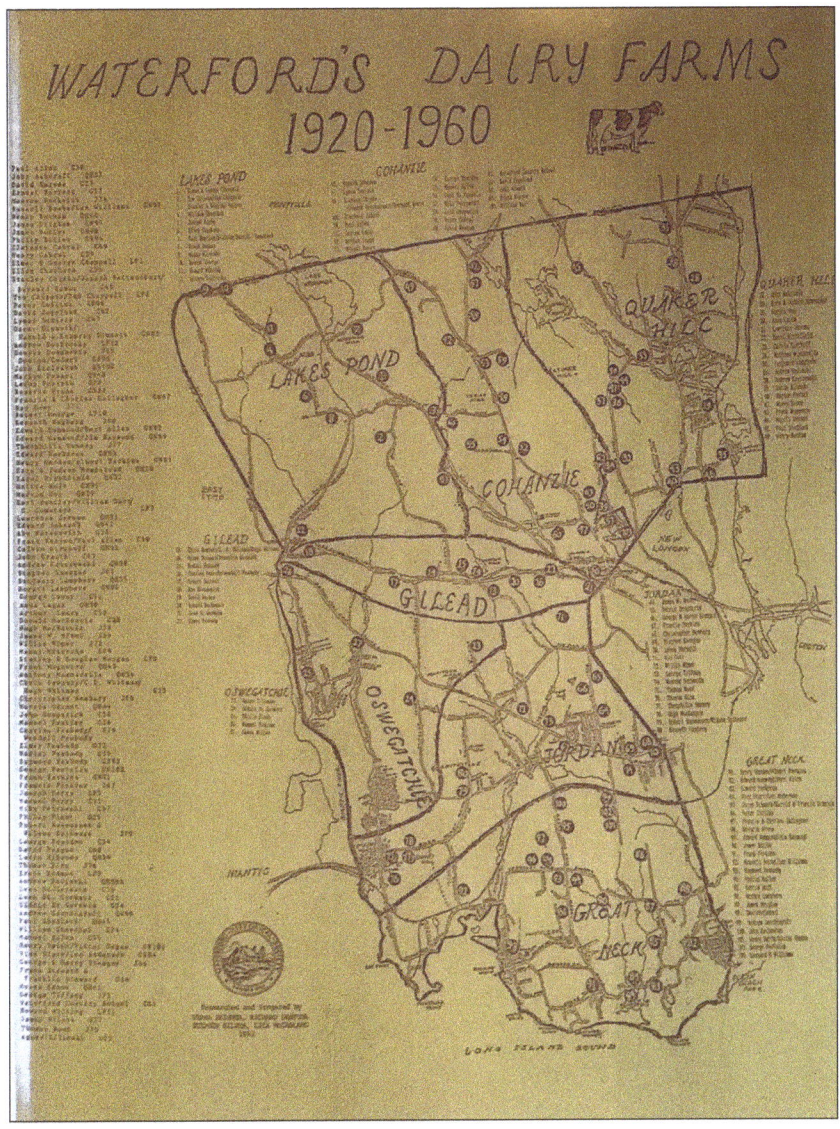

Figure 55 - Waterford Farms from the 1920s to '60s.

This is a map showing the location of 104 farms that existed in the town of Waterford between 1920 and 1960. There are very few of them left in 2021.

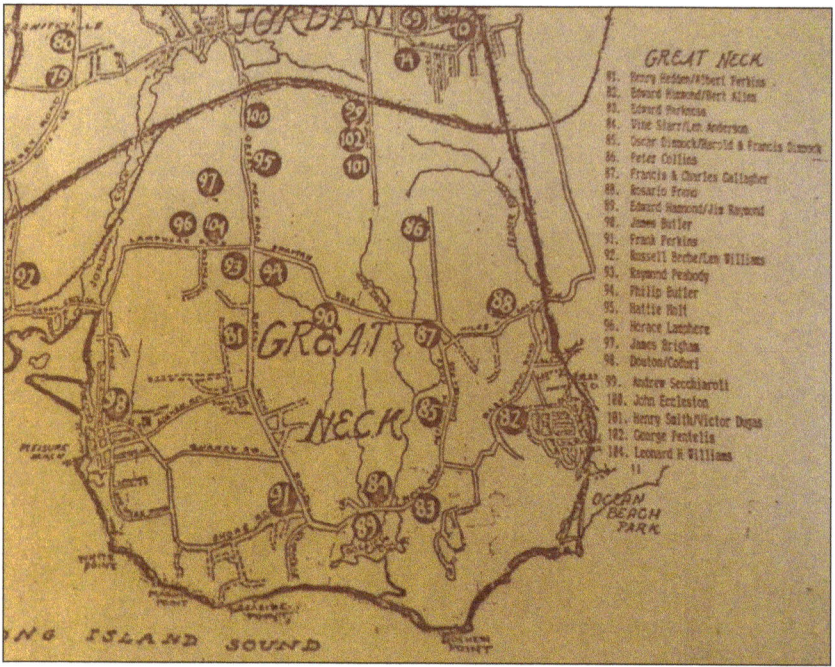

Figure 56 - The Great Neck portion of the Town of Waterford.

The detail of the map shown above shows the Great Neck section of Waterford, including our farm and the several farms around us. Our farm is Number 85 and occupied most of the property where the Great Neck label is shown. Number 87 is the Gallagher Farm, number 83 is Harkness and 82 is Hammonds, three of the local pastures that we used. Between the 85 and 82 labels was the Jacobs property that we used as a pasture. Number 84 is the Anderson farm.

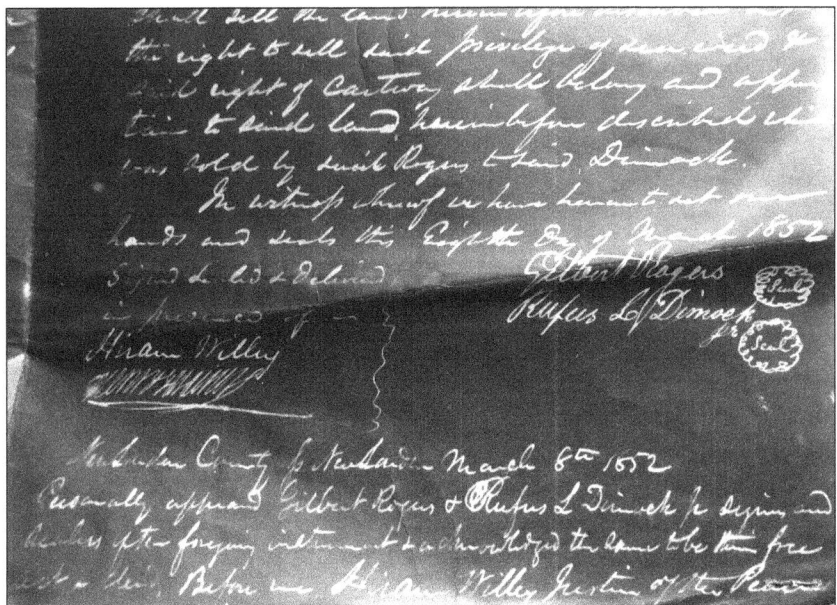

Figure 57 - Documentation of Seaweed Right.

The picture above is a copy of a portion of the Town of Waterford land records. While it's not very clear, it represents that the seaweed rights still in place today were confirmed on March 8, 1852 to have been a part of the original sale of the property, which happened in 1847. The document was signed by Gilbert Rogers, seller, and Rufus L. Dimock, buyer and my great-great-grandfather. Also, of note is that he spelled his name with one 'm' while we now spell it with two 'm's.

The significance of the seaweed right is that our family assumed that the right to transport seaweed from Common Beach to the farm was included in the original 1847 purchase of the property. It's not clear from the document what caused the controversy, but it was important that it was resolved as seaweed was important as insulation in the walls of the icehouse.

One can only surmise 170 years later that our ancestors assumed the seaweed rights came with the property when they bought it in 1847 but others at the time thought the rights were not included. Perhaps it had to do with how the seaweed was harvested at the beach, or how it was transported to our farm.

The document also clarifies that the original purchase was for 75 acres. At some point, additional land was purchased to total 114 acres. It is not clear to me, but it's possible the extra land was from the Anderson farm. The size would have been about right, and I do not recall Dad or Ike ever talking about that property as being anything but our property.

My sister Dot recently showed me a document that showed several sales of property from the Dimmocks to others. One such sale was the property to Booth Brothers Quarry. This suggests that the total property was more than 114 acres at certain times after 1852.

REFERENCES

Listed below are several YouTube videos that show details of many of the activities discussed in this book. As they say, a picture (or video) is worth a thousand words. A feature of YouTube is that, as you are watching one video, several other similar ones are recommended. This is highly informative but can also result in spending far more time than originally planned.

1. **"Horse drawn hay rake and hay loader"**
   (2:51 mins) – Larry Engle

   Comment: Good examples of both raking and the hay loader.

2. **"Amish hoist loose hay up into mow in barn"**
   (1:30 mins) – PineappleXVI

   Comment: Shows horses moving trailer into barn and lifting bundles of hay. Interesting how they swing the bundle before dumping it into the mow.

3. **"Unloading hay into the mow using horses and a hay hook"**
   (8:40 mins) – Larry Engle

   Comment: Interesting but takes a little too long to show various activities.

4. **"Beamish Pike maker demo"**
   (4:38 mins) – Bill Silloth

   Comment: Exceptionally good views of hay loader operation. Also shows filling of haystack frame and then flipping and releasing the frame to result in a haystack.

5. **"Horse drawn hay loader"**
   (1:140 mins) – Richard Hicks

   Comment: Good views of hay loader operation and how labor intensive it is on the trailer.

6. **"Rebuilding John Deere hay loader"**
   (7:06 mins) – Clover Martin Dairy

   Comment: Good description of the steps involved in rebuilding an old piece of machinery.

7. **"Reviving a hay trolley"**
   (6:59 mins) – Cozy Cow Family Farm

   Comment: Shows how the hay is lifted to the trolley and then travels to the desired spot in the mow. Details of repairs may be more than you are interested in.

8. **"Hay Trolley – how does it work?"**
   (4:03 mins) – Cozy Cow Family Farm

   Comment: Shows a hayfork being lifted to the rail, latching into the trolley & traveling to the desired release point.

List of Photos and Paintings

Cover – Mom's painting of the two barns
Dedication – Mom and Dad - p.ix
About the Author - Ray Dimmock - p.92

---

1. The Farm House, Hay Barn and Cow Barn. The Milk House is to the far left of the Farm House - p.xv
2. The Farm as viewed from the first hill -p.1
3. The bridge across the first brook, in the back half of the night pasture - p.2
4. Uncle Bert Bray picture of the barns - p.3
5. The gate leading to the plain, the first field mowed each year - p.5
6. Our Plow. The circular disks would slice the ground to start the furrow - p.6
7. The Mansion at Harkness State Park - p.8
8. The Mansion at Hammonds - p.9
9. Moving the young heifers to an on-site pasture - p.11
10. Uncle Ike Mowing with the C Tractor - p.13
11. Side Delivery Rake - p.14
12. Hay Loader - p.15
13. Blower - p.18
14. Raking with the H tractor - p.21
15. Haying with horses before the hay loader - p.22
16. Henry baling hay and carrying a rider sitting on the baling twine distribution buckets - p.24
17. Dad doing the baling, also with a rider sitting on the twine distribution containers - p.25
18. The '42 almost full - p.26
19. Hay barn from the back side prior to its demolition - p.28
20. The hay barn at Hammonds today, a prominent part of the Eugene O'Neil Theatre Centre - p.29

21. Horse-drawn mower - p.30
22. Trip rake - p.31
23. Henry and a helper throwing hay manually - p.31
24. Mom's painting of Dad driving the baler out to the plain - p.32
25. The pickup truck transporting cans of milk to the milk house - p.39
26. Cows going out after milking - p.41
27. Cows in the night pasture - p. 42
28. David feeding the young heifers - p.42
29. The bottler, filling bottles and installing caps - p.44
30. Milk stored in the refrigerator, 12 quarts per case, 18 cases per dolly - p.45
31. Dad returning from the day's deliveries - p.48
32. We gave a calendar to each customer every year - p.51
33. A sad note from one customer - p.52
34. This is the notice of business termination of Dimmock's Dairy - p.53
35. Mom working in one of her gardens - p.59
36. Prince & Chubby pulling the sled for a ride in the snow - p.61
37. Ray, Carol, John Starr and Mike Guilfoyle getting ready for 4H Fair, Uncle Ike in background - p.63
38. Ed, Dick, Dot and another child riding Pete - p.64
39. Kids going sledding the easy way - p.65
40. Ed and Dick feeding the chickens - p.66
41. Dad giving a ride to Al's wife Judy and their daughters Christina and Andrea - p.67
42. Bowles Barn after collapse of its first floor - p.67
43. View of second hill seen from the back of the plain - p.68
44. View of the farm from just before the second hill - p.69
45. Mom's painting of the rocks of Common Beach - p.70
46. Common Beach - Recent photo to compare with Mom's painting - p.71
47. Common Beach - Recent photo showing 'Big Rock' on left side - p.71
48. The creek between Ocean Beach and Waterford Town Beach - p.72
49. Public Notice sign between Ocean Beach and Waterford Town Beach - p.72
50. Wineberry at Bluff Point - p.73
51. Farm Layout - p.77
52. View of the farm from the first hill - p.78
53. Hay Fork - p.79

54. Son Jeff and granddaughter Delaney sitting on the C Tractor - p.80
55. A plaque on the wall at the Waterford Museum - p.81
54. Family picture at a recent 4th of July picnic, probably 2019 - p.82
55. Waterford Farms in the 1920s to 60s - p.83
56. The Great Neck portion of the Town of Waterford - p.84
57. Documentation of Seaweed Right - p.85

About the Author

Ray Dimmock

Ray, was born in 1946 in New London, CT. His family owned and operated a dairy farm, located in Waterford, CT and delivered milk to many homes in Waterford and New London. He was one of eight children in the family, with four older siblings and three younger. All of them worked on the farm in various capacities. At the age of 18, he went on to college and eventually to a 46-year career as an engineer at jet engine manufacturer Pratt & Whitney, headquartered in East Hartford, CT.

Ray is married, has three children and 10 grandchildren and now lives in Ashford, CT. He coached his kids in various sports highlighted best perhaps as coach of the Coventry High School girls' Basketball team for 12 years. A key milestone in that endeavor was a state championship during the 1987-88 season. For a few years, he also coached a ladies' softball team. That team achieved a league championship in one of those years.

Ray is now retired but volunteers at the US Coast Guard Academy and spends his leisure time enjoying family, traveling and reading. This book is his first but has inspired an interest in more.

www.ingramcontent.com/pod-product-compliance
Lightning Source LLC
Chambersburg PA
CBHW062040290426
44109CB00026B/2686